S216
Science: a Level Two Course

Environmental Science

TOPICS 5 to 8

Block 6

Cover image SeaWiFS image of the *global biosphere*

The Open University, Walton Hall, Milton Keynes, MK7 6AA

First published 2002. Reprinted 2006.

Edited, designed and typeset by The Open University.

Printed in the United Kingdom at the University Press, Cambridge.

ISBN 0 7492 6992 8

This publication forms part of an Open University course, S216 *Environmental Science*. Details of this and other Open University courses can be obtained from the Course Information and Advice Centre, PO Box 724, The Open University, Milton Keynes MK7 6ZS, United Kingdom: tel. +44 (0)1908 653231, e-mail ces-gen@open.ac.uk

Alternatively, you may visit the Open University website at http://www.open.ac.uk where you can learn more about the wide range of courses and packs offered at all levels by The Open University.

To purchase this publication or other components of Open University courses, contact Open University Worldwide Ltd, The Open University, Walton Hall, Milton Keynes MK7 6AA, United Kingdom: tel. +44 (0)1908 858785; fax +44 (0)1908 858787; e-mail ouwenq@open.ac.uk; website http://www.ouw.co.uk

2.1

s216 block 6i2.1

TOPIC 5

OCEANS and CLIMATE

Mark Brandon

Introduction

1

In this text there are references to 'The Earth's surface'. This is the poster map that accompanied Block 4, Part 1. You may find it useful to have the map to hand as you read this Topic so you can refer to it as and when indicated in the text.

When you look at a map of the Earth, it is obvious that the oceans cover a large part of the planet. However, a problem with all maps, particularly ones covering a large area, is that these representations of the Earth are a distortion of the truth because the Earth is not flat. It is quite a challenge to represent the distribution of features on the surface of a sphere on a flat surface. A better way of showing the extent to which the oceans cover the planet is to use a map projection of the Earth, as it would appear from space (but without the clouds). The projection in Figure 1.1 shows the largest ocean, the Pacific Ocean, from an altitude of 29 000 km. You may like to compare Figure 1.1 with your poster map 'The Earth's surface'.

You can see that virtually the whole field of view is water. Such a vast ocean puts into perspective the exploration and colonization of the Pacific island chains by the ancient peoples.

The oceans cover more than 70% of the Earth's surface, an area of almost 361 100 000 km^2. In this text we will investigate the huge and sometimes surprising effect this vast area of water has on the Earth's climate.

The average depth of the oceans is approximately 3.7 km, but there are regions where the ocean is over 10 km deep, and large areas where it is relatively shallow. In some places the oceans are deeper than the tallest mountains on the Earth are high. However, when we consider that the equatorial radius of the Earth is approximately 6380 km, it is apparent that the oceans are no more than a thin layer of water on the surface of the planet. But this thin layer contains almost $1.37 \times 10^9\ km^3$ of seawater.

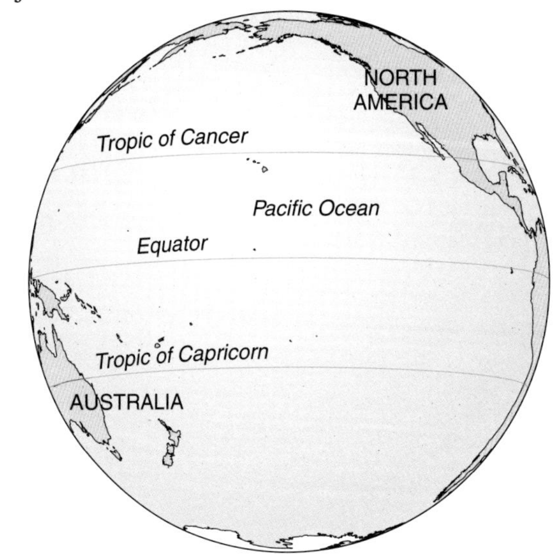

Figure 1.1 The way we would see the Pacific Ocean from an altitude of 29 000 km, but without the clouds.

1.1 The effect of the ocean on climate

We will take a look at the annual cycle of monthly air temperatures at two coastal locations that are at *different* latitudes to demonstrate the effect of this thin layer of water on the Earth.

- ❍ How would you expect the atmospheric temperature of a region to vary with distance from the Equator?
- ● You would expect the atmospheric temperature to fall with distance from the Equator.

This is, of course, the reason why many people like to go on their summer holidays to places located at relatively low latitudes such as the Caribbean rather than to colder areas such as northern Europe. But is this the whole story?

The annual atmospheric temperature cycles at Bergen in southern Norway (60° N, 5° E) and Halifax in Nova Scotia (44° N, 63° W) are shown in Figure 1.2. The relative locations of these two cities are depicted in Figure 1.3.

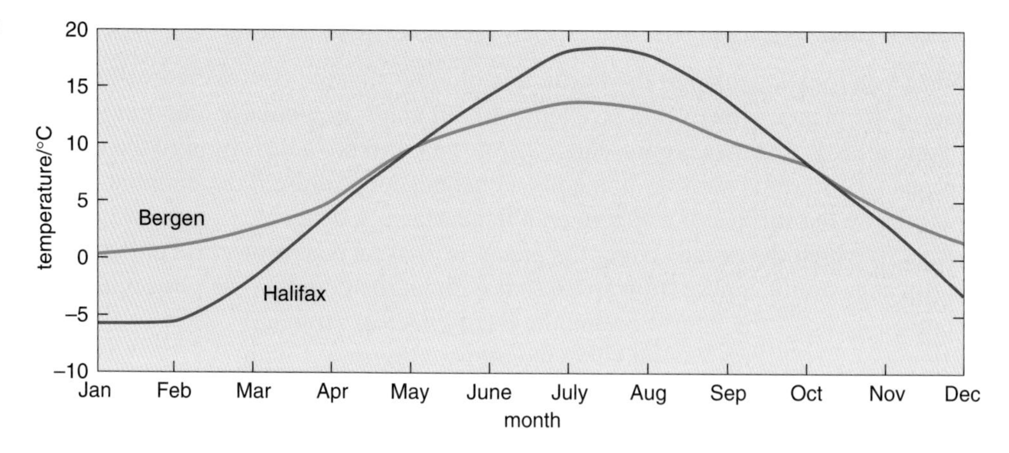

Figure 1.2 The monthly mean atmospheric temperatures for Bergen, Norway and Halifax, Nova Scotia.

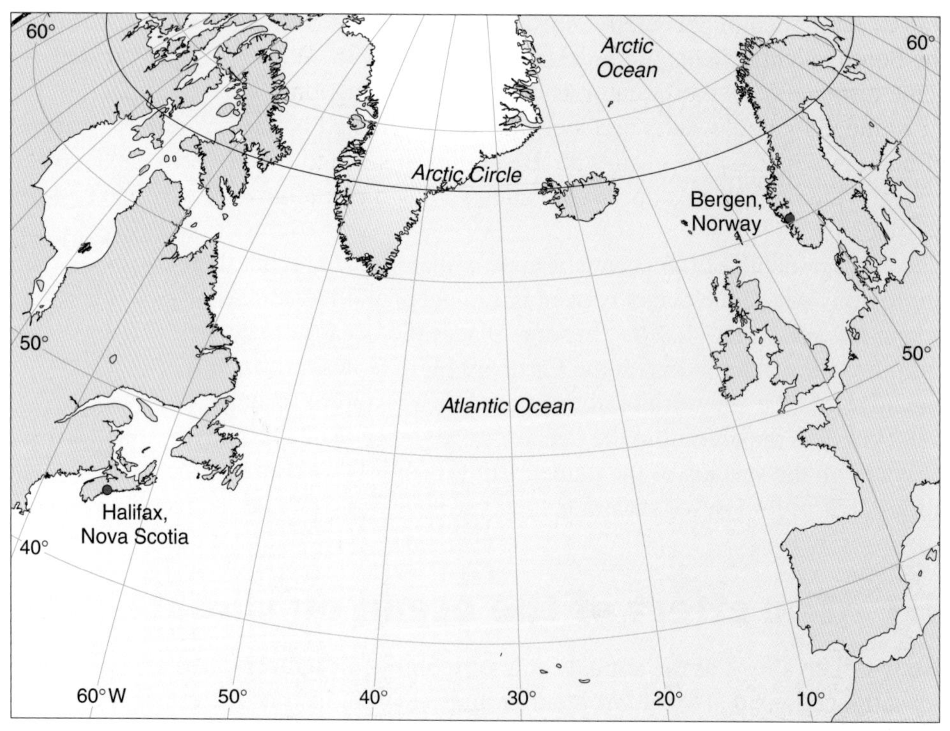

Figure 1.3 The relative locations of Bergen, Norway and Halifax, Nova Scotia.

Located at 44° N, Halifax is over 1700 km closer to the Equator than Bergen. For comparison, this is approximately the same distance from London to the Algerian coast in North Africa. Based on their locations, you would expect Halifax to be much warmer than Bergen. You can see from Figure 1.2 that this is not the case.

- What are the main differences between the atmospheric temperature cycles of Halifax and Bergen?

- Although Halifax is warmer in the summer, it is up to 6 °C colder than Bergen in winter.

This is rather surprising when you consider that Halifax is much further south than Bergen. You can also see that the range between the maximum summer temperature and the minimum winter temperature is also different between the two locations. In Halifax the range of temperature is approximately 24 °C, (about −6 °C to +18 °C) whereas in Bergen the range is approximately 12 °C (about 0 °C to +12 °C). From October to the end of April it is colder in Halifax than in Bergen and it is only from May to the end of September that Halifax is actually warmer than Bergen, despite being over 1700 km closer to the Equator. Clearly something rather strange is happening here — either Halifax is being cooled down in the winter months or Bergen is being warmed up. Although both these things are happening to a small extent, the main reason why Bergen is warmer than Halifax in the winter months is because a vast quantity of heat is being supplied to Bergen from the ocean. This heat comes from a remnant of a vast flow of warm water, called the Gulf Stream, which flows across the Atlantic Ocean from the Straits of Florida towards northern Europe. This flow of warm water moves a huge quantity of heat from low latitudes towards higher latitudes. The amount of heat that it transports is so huge that it swamps the cooling effect that we would expect from the increase in distance away from the Equator. For this reason we cannot just rely on latitude as an effective indicator of climate. An even more extreme example of this effect is seen on the island of South Georgia (Figure 1.4), which is located in the South Atlantic Ocean. This island lies at 54° S, 36° W and is approximately the same distance from the Equator as is the United Kingdom. South Georgia is surrounded by cold waters and icebergs that originated in the Antarctic. So instead of being warmed by the ocean like Bergen is, this island is subject to the opposite effect — the ocean cools down the island so much that almost 60% of it is covered by glaciers.

Figure 1.4 The coast of South Georgia in the South Atlantic Ocean.

In this brief introduction, you have seen how the oceans can influence regional climate and even turn conventional thinking upside down. In the next four sections we will investigate the reasons for this in more detail. First we take a look at ways of mapping the ocean basins. We also look at the shapes of the ocean basins because it is their shapes that are largely responsible for the patterns of water circulation in the global ocean. We will then investigate the properties of seawater, and the temperature, salinity and density structures in the Atlantic Ocean that arise because of the shape of the ocean basins.

We will discuss the forces that drive the global ocean's circulation and influence the climate of Bergen so much. Finally, we look at the implications for our climate of a change in the circulation of the global ocean.

1.2 Summary of Section 1

1 The oceans cover more than 70% of the Earth, and in places are deeper than the tallest mountains on the Earth are high. However, they are just a thin skin on the Earth's surface.

2 Heat can be transported by the oceans and this can affect regional climate.

Learning outcomes for Section 1

After working through this section you should be able to:

1.1 Recognize that the oceans cover most of the planet.

1.2 Describe how warm ocean currents can influence regional climate.

The shape of the oceans

2

The water in the oceans absorbs electromagnetic radiation so we can see very little of what is beneath the surface. So what is the sea floor like? If we could take away the water it would be easy to see what lies beneath, but of course we cannot do that. In fact, we know more about the shape of the surface of the Moon than the sea floor beneath the oceans. The enormity of this statement is brought home by the American journalist Robert Kunzig who pointed out that the American satellite *Magellan* mapped the permanently cloud-covered surface of Venus in 1992 to a higher resolution than we have mapped 90% of our oceans.

So let's take a look at what is known about the sea floor. If you look at the oceans on your poster map 'The Earth's surface' you can see there is a lot of variation in the shades of blue. Areas of light blue indicate shallow water and lines of very dark blue indicate very deep water. There are many shades of blue between these two extremes. So the depth of the global ocean is not uniform. In this section we look at various methods of measuring the depths of the ocean. We then examine the three large ocean basins of the Earth in detail.

2.1 Plumbing the depths

Until the late 1930s the only way to measure the depth of the ocean was to use a line with a weight on the end. Not surprisingly this took a lot of time. A ship would have to stop and lower a weight on a line to make a measurement (called 'taking a sounding'). In the early days this was done by hand, and when the line went slack the sea floor was judged to have been reached. There were many problems with this method, the most important being that the line used had a mass of its own. The mass of the line would soon become greater than the mass of the weight as line was paid out. This made it very difficult to decide when the weight had struck the sea floor. Such difficulties (combined with a lack of interest in knowing the depths of the ocean away from the navigational dangers of shallow waters near to the coasts) meant that measurements of the depth of the sea floor were few before the 19th century. One of the more memorable attempts to measure the depths took place in 1521 when Ferdinand Magellan stopped his ships in the Pacific Ocean during his famous circumnavigation of the globe. After paying out all of his line he was sure that the weight still had not touched the bottom. He obviously knew that the ocean was deeper than the line paid out — deeper than 400 fathoms, which is about 730 m — but then he mysteriously concluded that he was over the deepest depths in the oceans! We now know that this was definitely not the case.

Techniques progressed during the 19th century and a steam-powered winch was used to lower the sounding weight. This technique involved less physical effort. However, the ship still had to stop to take a sounding. The method of detecting when the weight touched the sea floor was also refined, but the actual lowering of the weight still took a long time.

Box 2.1 Viscosity of fluids

Any fluid, such as seawater or treacle, has a physical characteristic called **viscosity**. The greater the viscosity of a fluid, the longer it will take for something to fall through it under the influence of gravity.

- ❍ If two ball bearings are released simultaneously at the top of two 1 m-high jars, containing water and treacle respectively, which one will reach the bottom first, and so which fluid has the greater viscosity?
- ● The ball bearing in the jar containing water will reach the bottom much quicker than the ball bearing in the jar containing treacle, so the treacle has the greater viscosity.

The upper limit of the speed at which something sinks freely under gravity in a fluid is given by a mathematical equation called Stokes' law. This equation depends principally on the shape of the object and the viscosity of the fluid. In our 1 m-high jar containing water, the ball bearing will still be accelerating as it hits the bottom of the jar. However, in the jar containing the treacle the ball bearing will accelerate initially, it will then continue to fall at a constant speed of descent given by Stokes' law. If the water in the jar were hundreds of metres deep, the ball bearing in that jar would also reach a constant speed of descent given by Stokes' law.

The sinking of RMS *Titanic* in the North Atlantic Ocean in 1912 provides an excellent example of the length of time it takes to reach the sea floor.

Dr Robert Ballard, who discovered the wreck in 1985, concluded that, after the initial acceleration as the wreck left the surface, the speed of descent would have been in the range of 40–48 km per hour. This speed is limited by Stokes' law.

- ❍ If the *Titanic* sank at 40 km h^{-1} and the water depth was 3797 m, how long would it have taken for the wreck to reach the sea floor?
- ● The ocean is 3797 m deep and the wreck was descending at 40 km h^{-1}. This means that the time taken to reach the sea floor would have been 3797 m/(40 000 m h^{-1}), which is 0.095 hours. We can covert this to minutes: $0.095 \times 60 = 5.7$ minutes. So the *Titanic* would have taken 5.7 minutes to reach the sea floor.

For this calculation we have assumed that the ship reached the speed of 40 km h^{-1} as soon as it left the surface of the ocean. In reality, the ship would have sunk slowly at first and then accelerated to this constant speed. Once the ship had started to sink it would, of course, have gained momentum in a downward direction (momentum = mass × velocity). When the *Titanic* hit the sea floor the momentum was absorbed and the huge impact buried the ship to a depth of almost 20 m. Dr Ballard showed that it was this crash into the sea floor that actually shattered the *Titanic* rather than the initial collision with an iceberg. You may think that almost six minutes for the *Titanic* to reach the sea floor is a relatively long time — yet it did not even sink in a deep part of the ocean.

It would have taken a minimum of two hours, including the time taken to stop and restart the ship, to make one sounding in water of average depth (approximately 3700 m) using a steam-powered winch to lower and raise the sounding weight to the sea floor at a speed of 2 m s^{-1}. This is one of the reasons why measurements of the oceans depths were rarely carried out in the 19th century.

The first real impetus for finding out about the depths of the sea floor was, not surprisingly, commercial. In the 1850s our understanding of electricity had expanded to the point where telegraphic cables could be laid between various cities on the mainland. It was only natural that investors soon began to look at connecting the continents. Before long Great Britain was linked with France, and

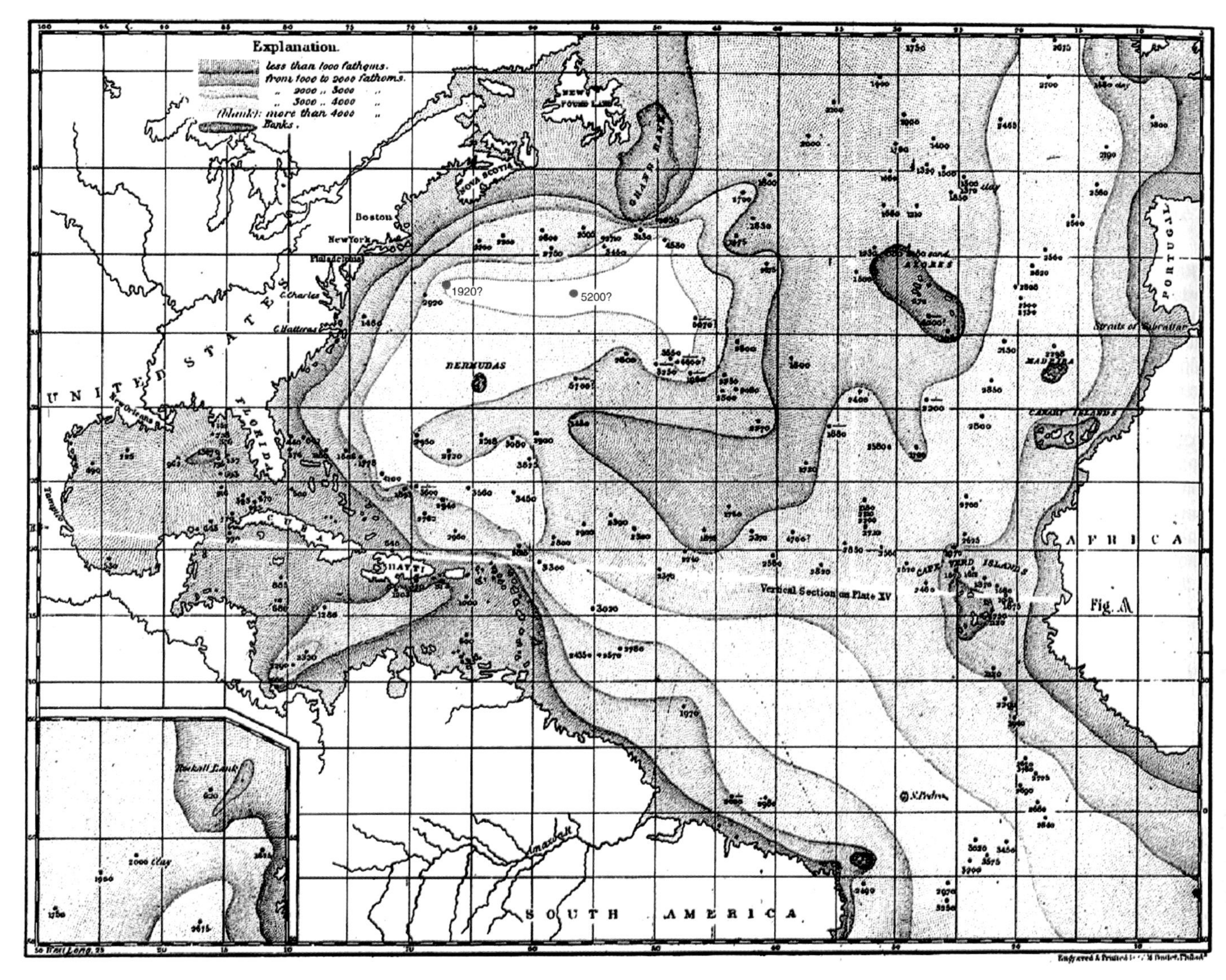

speculators looked at laying telegraphic cables across the Atlantic Ocean to connect North America with Europe. The depth of the oceans was absolutely critical in the budget for such a project: it helped to determine how much cable was required. The only map of the depths of the North Atlantic Ocean that was available at the time had been drawn by the American sailor Lieutenant Matthew Maury. Maury pooled all the measurements of the depths of this ocean that were available, along with some measurements he had made himself on a ship called the *Dolphin,* to draw the map in Figure 2.1.

Although Maury's map was only based on about 200 measurements, when you compare Figure 2.1 with the North Atlantic Ocean shown in the poster map it is actually not that poor a representation. Maury's map showed that the depth of the ocean rapidly increased away from the coasts. The shallower region extending to the southwest of the Azores was the first indication of the stunningly vast Mid-Atlantic Ridge sub-marine mountain chain.

Figure 2.1 Matthew Maury's map of the depths of the North Atlantic Ocean as it appeared in 1861. Note that even though he only had 200 data points he was suspicious enough of some data to place question marks next to two data points east of Philadelphia. These points have been enlarged and coloured red to make them clearer.

Box 2.2 The telegraphic plateau across the North Atlantic Ocean

When Matthew Maury drew his map of the North Atlantic Ocean he had no idea of how quickly it would become commercially important. The Atlantic Telegraph Company (ATC) had been developing plans for linking the continents, and were proposing a telegraphic link between Newfoundland and Ireland — a distance of over 3000 km and depths of almost 3000 m. Speculators had contacted Maury in the early 1850s to ask whether such a link would be feasible, and Maury replied:

> The bottom of the sea between the places is a plateau which seems to have been placed there for the purpose of holding the wires of a sub-marine telegraph.

This is a rather confident assertion on the basis of the map in Figure 2.1! The project was considered so important that the British and American governments provided ships to carry out the task. The Americans provided one of their finest ships: the USS *Niagara*, whilst the British supplied HMS *Agamemnon*, a decrepit ship built just after the Napoleonic War and close to retirement. Nevertheless, the ships managed the job and, in 1858, the ATC successfully laid the cable across numerous sub-marine mountain chains and valleys that Maury had never known about. It was a very expensive venture. Money had been raised from both governments and speculators. Unfortunately this first link failed after about three months. However, the first message passed by the British government between the UK and the USA — giving a British army regiment orders *not* to ship to India to fight in the Indian Mutiny as the conflict was over — saved a fortune and proved the worth of rapid communications. The two ships returned in the following years to successfully complete the link.

Over the following century sounding machines became more sophisticated and efficient. These modifications enabled measurements to be taken at a faster rate and samples of the sea floor to be collected. Despite such modifications the process was still basically the same: a weight was lowered to the sea floor, and then recovered.

2.2 The entrance of acoustics

The next leap forward in mapping the ocean floor was based on acoustics. During the First World War sound was used to detect submarines. This technique works in the following way:

- A sound pulse is emitted at the ocean surface.
- It travels through the seawater to the seabed and is reflected back to the surface, where it is detected by a receiver.
- The time delay from the initial sound pulse to its reception is measured.
- The depth is calculated using the speed of sound in seawater (Figure 2.2).

This technique became known as Sonar (*SO*und *N*avigation *A*nd *R*anging) and it changed the mapping of the deep forever.

Question 2.1

A ship transmits a pulse of sound to the sea floor and detects the echo 6 seconds later. Given that the speed of sound in seawater is $1500\,\mathrm{m\,s^{-1}}$, what is the depth of the sea floor at this point?

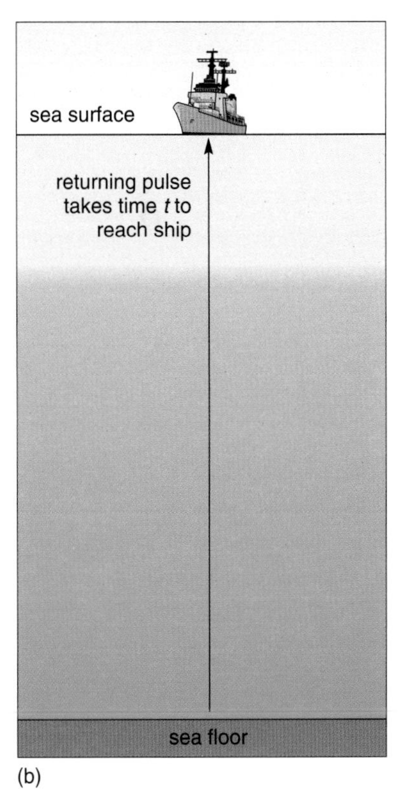

Figure 2.2 (a) Ship transmitting a pulse of sound that takes time t to reach the sea floor. (b) The ship receiving the echo from the sea floor. The total time for the echo to reach the ship after transmission is $2t$ seconds. The depth can be derived using the speed of sound in seawater.

Each one of the measurements Maury used to create Figure 2.1 could take several hours, depending on the sounding equipment used. But using acoustic techniques each sounding only takes a few seconds. In fact it was such a leap forward that when the USS *Stewart* made the first continuous crossing of the North Atlantic using a sonar system in 1922, in the 9 days it took to cross the Atlantic, 900 soundings of the depth of the sea floor were made. This was almost 5 times the data set that Maury had used to draw his map.

Soon many ships were traversing the oceans and making acoustic soundings of depths. It was not long before the British survey ship HMS *Challenger* discovered the Marianas Trench in the North Pacific Ocean at a depth of over 10 800 m. Sound took almost 14 seconds to travel to the sea floor and back again in this depth of water and the strength of the echo reflected back from the sea floor was difficult to detect using normal hearing. It took HMS *Challenger* over 6 hours to check the measurement using a conventional steam-powered sounding machine. As the database of ocean depths increased new scientific thinking was required to interpret the data.

The majority of ships traversing the oceans were then, as now, commercial. They sailed the same sections of ocean between the major global ports. Naturally, this left large areas of the oceans that had not been surveyed. The regions that were particularly deficient in measurements were those distant from major habitation, such as the Antarctic and the South Indian Ocean. Some of these gaps in ocean-depth data were filled in by the next jump forward in mapping the deep — the use of a technique called satellite altimetry.

2.3 The future reaches the ocean depths

At first it may seem unlikely that satellites can be used to map the sea floor because electromagnetic radiation does not pass through seawater. Nevertheless, ocean-floor topography can be obtained from space. The height of a satellite above the surface of the Earth can be measured, using radar, to an accuracy of within a few centimetres. The level of the ocean surface is influenced by several factors including the wind, the tides and the mean circulation. If these factors are removed by averaging the data to give a 'mean' picture, then it can be seen that the ocean surface is not level, but follows what is called the **geoid**. The geoid is the surface of equal gravitational energy. It has vertical variations of up to several metres over horizontal scales of hundreds of kilometres. These reflect the local gravity field, which is related to the shape of the sea floor. This is because a mountain in the ocean (called a seamount) has increased local gravity and so attracts water over the top of it. This creates a bulge at the sea-surface that can be measured by a satellite. This principle of satellite altimetry can be seen in Figure 2.3.

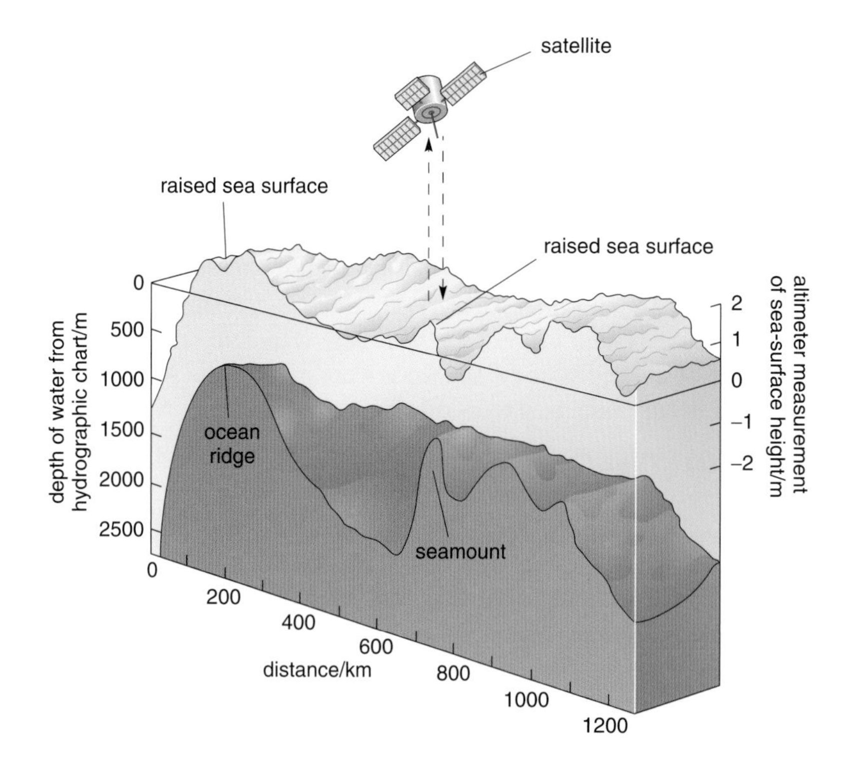

Figure 2.3 The principle of satellite altimetry. Local variations in the shape of the sea floor are reflected in the mean shape of the sea-surface. By careful processing, the satellite data can be used to map the sea floor.

By careful processing of satellite data and calibration in regions where the ocean has been well surveyed, satellite altimetry can be used to map areas of the ocean that have never been visited by ships to give what is called 'predicted bathymetry'. Two American scientists, Sandwell and Smith, used this technique to produce the sea-floor data used in your poster map 'The Earth's surface'.

❍ Can satellite altimetry be used to tell us about areas of the ocean that are covered in sea ice?

● No. The ice cover is constantly moving and has a variable height above the sea-surface. It is not possible to remove the signal from the ice from the altimetry data and so the shape of the sea-surface cannot be determined using this technique.

This means that the areas of the world that are permanently frozen and covered in sea ice, such as the Arctic and Antarctic, still have to be mapped by ships and drifting ice stations using sonar, and where there is no data available a 'best guess' is constructed based on the available evidence. Figure 2.4 shows an example of this in a region of the western Weddell Sea in Antarctica. Virtually all of the area in this Figure is partially covered in sea ice throughout the year, and there is permanent ice cover in the region bounded by 67° S, 48° W, in the bottom left of Figure 2.4. A scientific research station called Ice Station Weddell drifted through this region, indicated by the red line, in 1992. This ice station provided virtually the only actual data within the permanent ice-covered region and yet the chart still shows some bathymetric contours away from this red line. This demonstrates that even today we do not know how deep some regions of the ocean are.

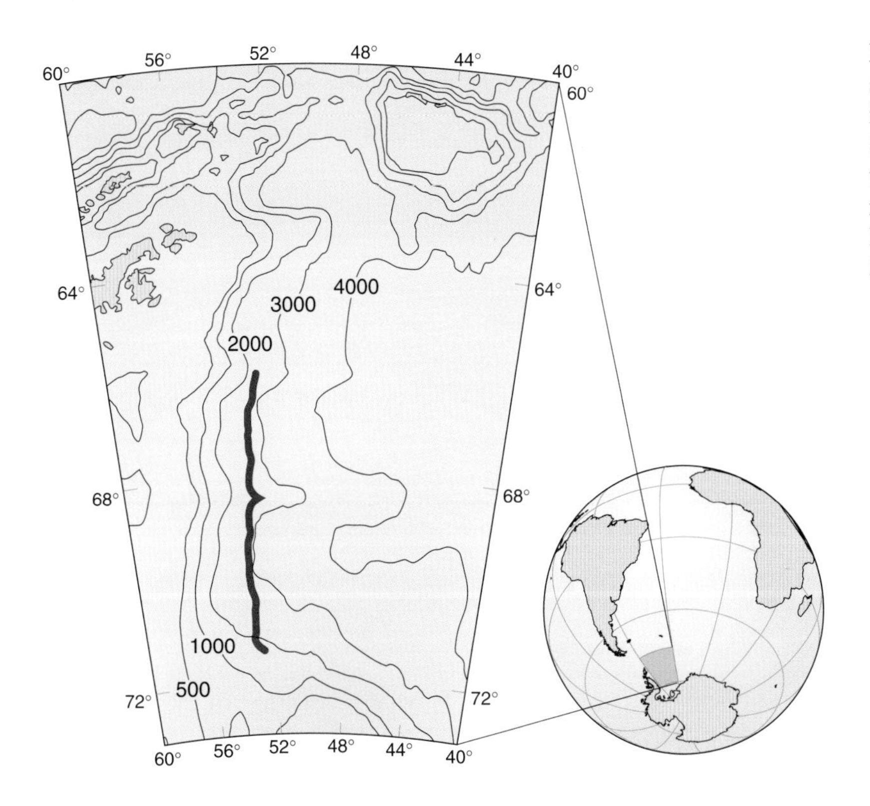

Figure 2.4 A map of the western Weddell Sea showing the bathymetric contours. The red line is the track of a drifting ice station called Ice Station Weddell. This is the only place where there is any data on the ocean depths in the region bounded by 67° S, 48° W and the bottom left of the Figure. Contours on the figure are at 500 m, 1000 m, 2000 m, 3000 m and 4000 m.

The technique that Sandwell and Smith used to create a data set has a horizontal resolution determined by the satellite's orbit. At low latitudes the resolution can be as good as 1 km, but as latitudes get higher the resolution decreases to approximately 10 km. The amount of detail decreases as the horizontal resolution decreases. Imagine what a contour map of the UK would look like if you only had one average height in each 10 km square — many hills and valleys would disappear from the map based on this data.

In measuring the depths of the ocean, along with so many things, the detail is the most problematic aspect. In practice this means that the use of acoustics to measure the ocean depths still has an important part to play in mapping the deep. It is the only technique that can easily give a large amount of detail.

2.4 The ocean floor

Individual features of the Earth's surface and ocean floor are covered in Block 4, Part 1; here we consider the general features of the ocean floor. Figure 2.5 shows a special type of curve called a hypsographic curve. This shows the cumulative frequency of the topography on the Earth both above and beneath the surface of the oceans. To use this type of plot we look at how much each region takes up on the x-axis. For example the area of the Earth that is covered by continental slopes can be determined from Figure 2.5. On the x-axis the region of continental slope is bounded by the limits of approximately 35–55%. This means that about 20% of the Earth's surface is covered by continental slopes (55% – 35% = 20%). You can also see that the average depth of the oceans is 3.7 km, whereas the average height of land above 0 km altitude is much less at 0.8 km.

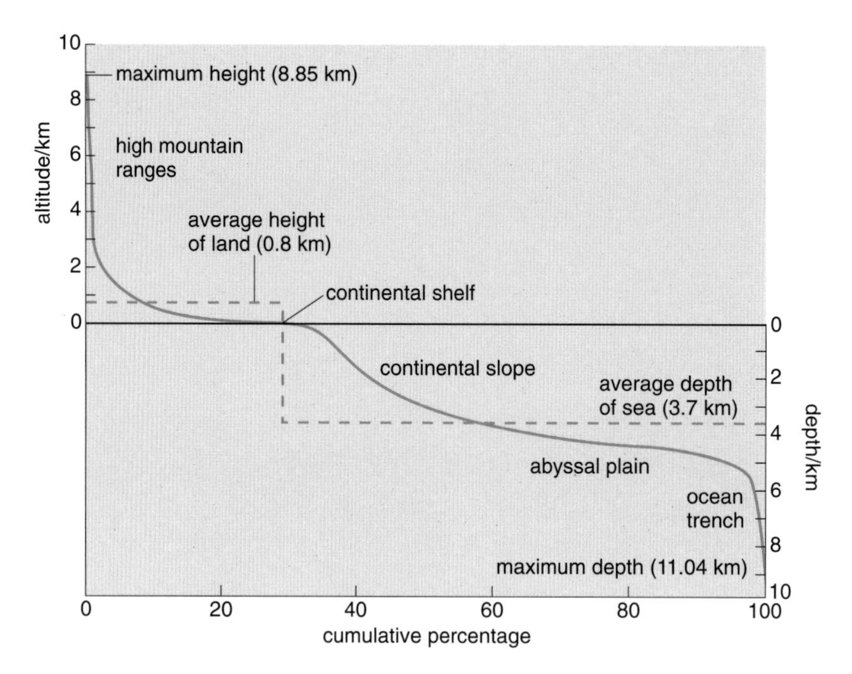

Figure 2.5 Hypsographic curve showing the cumulative frequency of the amount of the Earth's surface that lies above, below, or between given depth or height levels. The zero level on the *y*-axis is sea level.

❍ From the hypsographic curve in Figure 2.5, what percentage of the Earth's surface is covered by abyssal plains?

● Abyssal plains cover approximately 35% of the surface of the Earth.

The abyssal plains cover the greatest area of the Earth's surface. They are vast regions covered with thick layers of sediment. This makes them the flattest regions on the planet, with average gradients of less than 0.05°. We saw in Block 4, Part 1 that the gradient of the continental slope is approximately 4°. Most of the topography across the whole of the North Atlantic Ocean lies within these two gradients. This means that if we were able to pull the plug from the seas, and if we did not sink into the ooze, it would be quite easy actually to drive across the sea floor from Europe to North America with a 4×4 vehicle if, of course, we avoided the occasional vertical cliff caused by geological faulting. Even more surprising is that although, as Figure 2.5 shows, the ocean trenches can be over 11 km deep, they descend on one side from an abyssal plain that is already 5–6 km deep. If we take the width of the Marianas Trench as approximately 60 km, the gradients should probably be no greater than about 8%. So, if the water were removed, we could drive to the bottom of the trench in our 4×4! It is certainly not possible to drive to the top of the highest mountain. This demonstrates the comparative gradients of the oceans and continents.

Such examples of the gradients present in the oceans demonstrate the apparent vertical magnification in an image of the sea floor such as that shown in the poster map and Figure 2.5. It is hard to appreciate the true scale of ocean features. The Earth has a radius of approximately 6380 km at the Equator. Figure 2.5 shows that the mean ocean depth is 3.7 km. The relative thickness of the layer of water on the Earth's surface is about the same as the thickness of the layer of skin on an apple, and yet we have seen in Section 1 how much this thin water layer can affect regional climate.

One of the reasons why the ocean can affect the Earth's climate so much is the shape of the ocean basins.

2.5 The shape of the ocean basins

There are three large ocean basins on Earth. These are the Pacific Ocean, the Atlantic Ocean and the Indian Ocean. Their key characteristics are listed in Table 2.1 and shown in Figure 2.6. The bathymetric details within the individual basins can be seen on the poster map.

Table 2.1 Approximate areas of the principle ocean basins.

	Ocean			Global ocean
	Pacific	Atlantic	Indian	
ocean area/10^6 km^2	180	107	74	361
land area/10^6 km^2	19	69	13	101
average depth/m	3940	3310	3840	3730

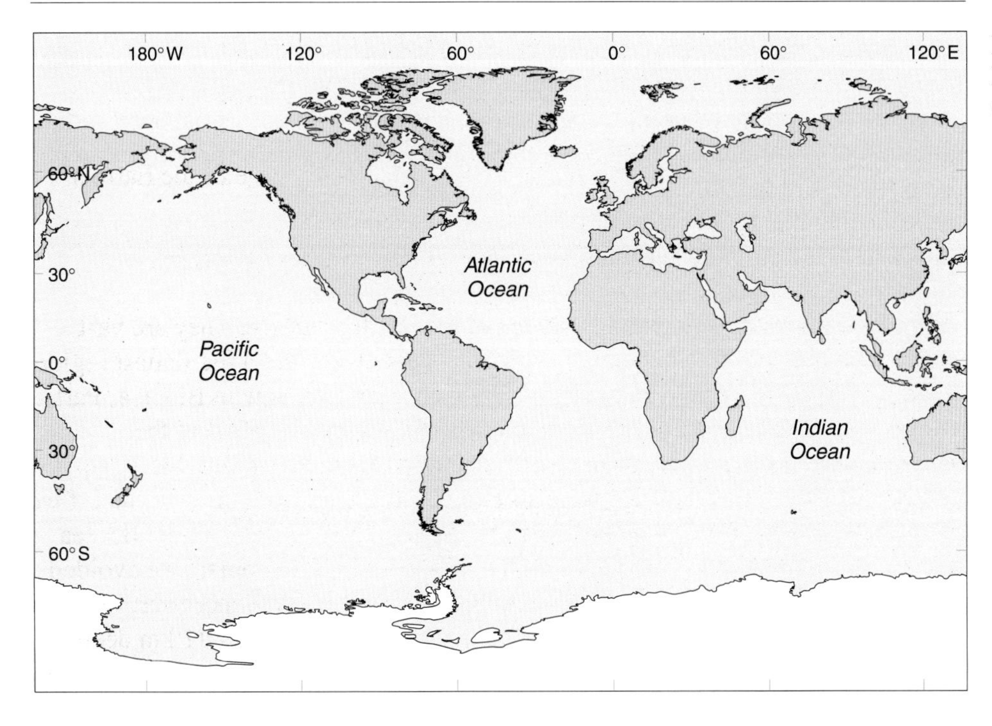

Figure 2.6 Projection of the Earth showing the three major ocean basins.

2.5.1 Pacific Ocean

The Pacific Ocean is, on average, the deepest ocean and has proportionally the least land area. The lack of land means that the wind flow across the ocean is relatively undisturbed. This undisturbed wind flow has a large effect on the resulting global ocean circulation. Despite being the largest ocean, the flow of water to latitudes north of 60° N is virtually blocked by the Asian and North American land-masses. There are deep ocean trenches on three sides of the basin. The ocean is virtually open to the southwest through the Indonesian Archipelago, and to the south — the only place where there are no deep ocean trenches.

2.5.2 Atlantic Ocean

This ocean basin has a much smaller east–west extent than the Pacific and Indian Oceans, and has the greatest proportion of land area. It also has a number of other key features that distinguish it, the most important being that it is almost completely open to both the north and the south. The total north to south extent of the ocean is over 20 000 km. It extends from the Arctic Ocean to the coast of Antarctica. Deep trenches occur only in the southwest corner of the South Atlantic. The Atlantic is divided almost equally into east and west basins by the huge Mid-Atlantic Ridge. This ridge rises from the abyssal plain to almost within 2000 m, and in places to within 1000 m, of the surface. This topography has a very large effect on the ocean circulation in the Atlantic. On the edges of the Atlantic Ocean are marginal seas, which include the Mediterranean Sea and the Arctic Sea. These two seas also have a strong influence on the circulation of the Atlantic, and the resulting global ocean. Fractures caused by transform faults (see Block 4, Part 1) within the Mid-Atlantic Ridge are also critical for the ocean circulation.

2.5.3 Indian Ocean

The Indian Ocean is the smallest of the large oceans, and is the least explored. It is triangular in shape and bounded by Africa to the west, India to the north and Australasia to the east. Despite being bounded to the east by land, there are many deep gaps between the islands that separate this ocean and the Pacific Ocean. There are no deep trenches and, apart from a large area of continental shelf to the northwest of Australia and Indonesia, the Indian Ocean has a fairly uniform depth. The ocean is divided into deep basins by a mid-ocean ridge that is much less extensive than the ridge system in the Atlantic Ocean. The Indian Ocean is very important in the global ocean circulation as it links the other two basins and is subject to very strong seasonal influence by the monsoons.

2.6 Summary of Section 2

1 Initially, the ocean depths were mapped by lowering a weight to the sea floor and taking measurements. Acoustics made the measurements much easier to take, but the real leap forward came with satellite technology.

2 There are still areas of the ocean where the bathymetry is not well known.

3 The average depth of the world's oceans is 3.7 km. The oceans are, on average, deeper than the Earth is high.

4 The three major ocean basins have distinct shapes. The largest, the Pacific Ocean, is closed to the north and has the smallest land area. It also contains virtually all of the ocean trenches. The Atlantic Ocean is open to the north and south and is split into two by an extensive mid-ocean ridge system. The third major ocean basin is the Indian Ocean.

Learning outcomes for Section 2

After working through this section you should be able to:

2.1 Describe different methods of measuring the depths of the global ocean.

2.2 Name areas of the oceans where knowledge of the sea floor may be poor or non-existent.

2.3 Outline the principle of satellite altimetry.

2.4 Interpret a hypsographic curve.

2.5 Describe the main differences between the three major ocean basins.

Seawater

3

As every mermaid knows the most obvious difference between seawater and freshwater is that seawater tastes salty! As well as changing the taste, the presence of dissolved salts in water radically changes its physical properties. In this section we will look at some of these properties. We will also look at the typical vertical distribution of temperature and salinity throughout the oceans.

3.1 Salt in the ocean

The saltiness of seawater is not just due to sodium chloride (Na^+ and Cl^- ions). It is a mixture of many different ionic constituents. If we placed 1 kg of seawater in a pan and boiled all the water away, we would find, on average, 34.482 g of solid material left in the pan. Almost 99.95% of the total mass of this solid material is made up of the constituents listed in Table 3.1. In reality we would end up with less salt than this as some of the ionic constituents react with the atmosphere to form gases — so this is not really a good way to see which salts are contained in seawater.

Table 3.1 Major ionic constituents dissolved in seawater.

Ion	% by mass of 34.482 g	Weight/g kg^{-1}
chloride, Cl^-	55.04	18.980
sulfate, SO_4^{2-}	7.68	2.649
hydrogen carbonate, HCO_3^-	0.41	0.140
bromide, Br^-	0.19	0.065
borate, $H_2BO_3^-$	0.07	0.026
fluoride, F^-	0.003	0.001
sodium, Na^+	30.61	10.556
magnesium, Mg^{2+}	3.69	1.272
calcium, Ca^{2+}	1.16	0.400
potassium, K^+	1.10	0.380
strontium, Sr^{2+}	0.04	0.013

One question we need to consider is whether the ionic constituents dissolved in seawater are the same in all the oceans of the world. HMS *Challenger* was the first ship completely devoted to science to sail around the world on a four-year voyage between 1872 and 1876. During this voyage, 77 seawater samples were collected and returned to Britain for analysis. By quantitative analysis of the salt content it was found that, although the *total* amount of dissolved salt varied from place to place, the major elements in seawater were always present *in the same relative proportions*. This amazing fact is called the principle of **constancy of composition**.

- ❍ If the ratio of the amount of K^+ ions to Cl^- ions in the Atlantic Ocean was 0.02, what would be the ratio of the two ions in the Pacific Ocean?

- ● The ratio of the K^+ ions to Cl^- ions would be exactly the same at 0.02 because of the constancy of composition of seawater.

The constancy of composition of seawater is very useful for oceanographers because just one of the ionic components listed in Table 3.1 needs to be measured to derive the total salinity of the seawater. Up until the late 1970s salinity was actually defined as being equal to 1.80655 times the amount of the Cl^- ion present.

There are 34.482 g of salt in 1 litre (1000 g) of seawater of average density. Salinity values used to be given in parts per thousand (abbreviated to ppt or given the symbol ‰). So in seawater of average ocean density the salinity would have been expressed as 34.482‰. In 1978 a new salinity scale was developed which was based on the conductivity of a seawater sample compared to the conductivity of a reference sample of 'standard seawater', as the conductivity of seawater is directly proportional to its salt content. The conductivity scale has no units since it is effectively describing a ratio, but for all practical purposes it is equivalent to ppt. For this reason we sometimes still see salinity values quoted in ppt or ‰ when really they should have no units. The salinity of seawater of average ocean density is 34.482.

Table 3.2 Percentages by mass of the most abundant elements in the Earth's continental crust.

Element	%
oxygen, O	46.6
silicon, Si	27.7
aluminium, Al	8.3
iron, Fe	5.0
calcium, Ca	3.6
sodium, Na	2.8
potassium, K	2.6
magnesium, Mg	2.1
titanium, Ti	0.4
other elements	0.9

3.2 Where does the salt in the oceans come from?

Analysis of marine evaporite deposits and other sediments in the oceans indicate that the ocean salt composition has not really varied in the last 10^8 years. Given that the residence time of water in the ocean is about 4000 years, the fact that the salt composition has not changed indicates that the processes that add salt to the ocean and the processes that remove salt from the ocean must be in balance. We can write this as:

$$\text{salt added to world ocean} - \text{salt removed from world ocean} = 0 \qquad (3.1)$$

So where is the salt coming from? If salt is removed from the ocean in evaporites and other sediments it must be added from somewhere else for the salinity to stay constant. Table 3.2 shows the percentages by mass of the most abundant elements in the Earth's continental crust.

❍ Which elements appear in both Table 3.1 and Table 3.2?

● The elements that appear in both Tables are calcium, potassium, magnesium and sodium.

These elements are all soluble products of chemical weathering and are washed down into the oceans. The most significant component for the salinity of the oceans is sodium (Table 3.1). The other major constituents by mass are chlorine and sulfur. These come from a different source. Table 3.3 shows the chemical composition of volcanic gases emitted by a present-day Hawaiian volcano.

Table 3.3 Chemical composition of the gases emitted by an active Hawaiian volcano.

Chemical constituent	% by number of molecules
water, H_2O	80.4
sulfur dioxide, SO_2	13.5
carbon dioxide, CO_2	3.7
hydrogen sulfide, H_2S	1.0
hydrogen, H_2	0.9
hydrogen chloride, HCl	0.2
hydrogen fluoride, HF	0.2
carbon monoxide, CO	0.1
nitrogen, N_2	trace
argon, Ar	trace

The chlorine and sulfur present in HCl, SO_2 and H_2S are soon washed out of the atmosphere by rain and are deposited in the oceans. Hence the components of the salts dissolved in seawater listed in Table 3.1 come from two very different sources — the chemical weathering of rocks, and volcanic activity. The processes by which salts are removed from the ocean by the precipitation of some solids, and the increase in ocean salinity from the combination of volcanic gases and weathering components, are in balance, as stated in Equation 3.1.

3.3 Freshwater and seawater: why is seawater so special?

Freshwater and seawater differ in their salinity and density. Both these physical properties vary with temperature.

Let's consider a swimming pool filled with freshwater. In winter the air temperature falls and the temperature of the water at the surface of the pool decreases. The density of freshwater against temperature is shown in Figure 3.1.

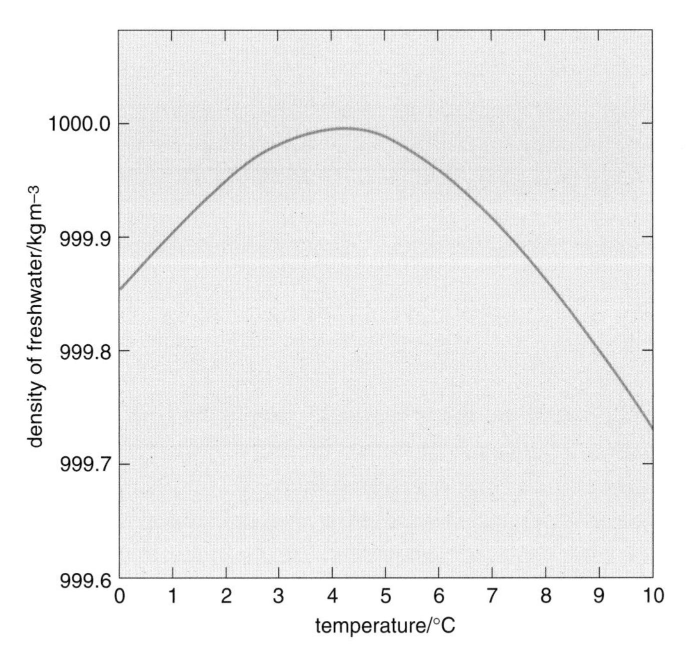

Figure 3.1 The density of freshwater against temperature.

- At what temperature is freshwater at its maximum density?
- From Figure 3.1 we can see that freshwater is at its maximum density at approximately 4 °C. At both colder and warmer temperatures the water is less dense.

Figure 3.1 shows that the surface water becomes more dense as it cools towards 4 °C. This cooler surface water is more dense than the warmer water beneath it so it sinks (through convection). The warmer water then rises to the surface of the pool. With continued cooling this warmer surface water also becomes more dense and it too sinks. During this phase when the surface water is being cooled the water in the swimming pool will become **stratified**, that is the density will vary with depth, with the less dense water at the surface. The density of the water in the swimming pool will increase rapidly with depth until the maximum density is reached, which is when the temperature is 3.98 °C. This process continues until all the water in the swimming pool has reached the temperature of maximum density.

Question 3.1

What will happen to the water in the swimming pool when all the water is cooled below 3.98 °C?

Eventually the surface layer of water will be cooled to the freezing point and ice will form at the surface. At this point the temperature and density of the water will have a structure like the one shown in Figure 3.2.

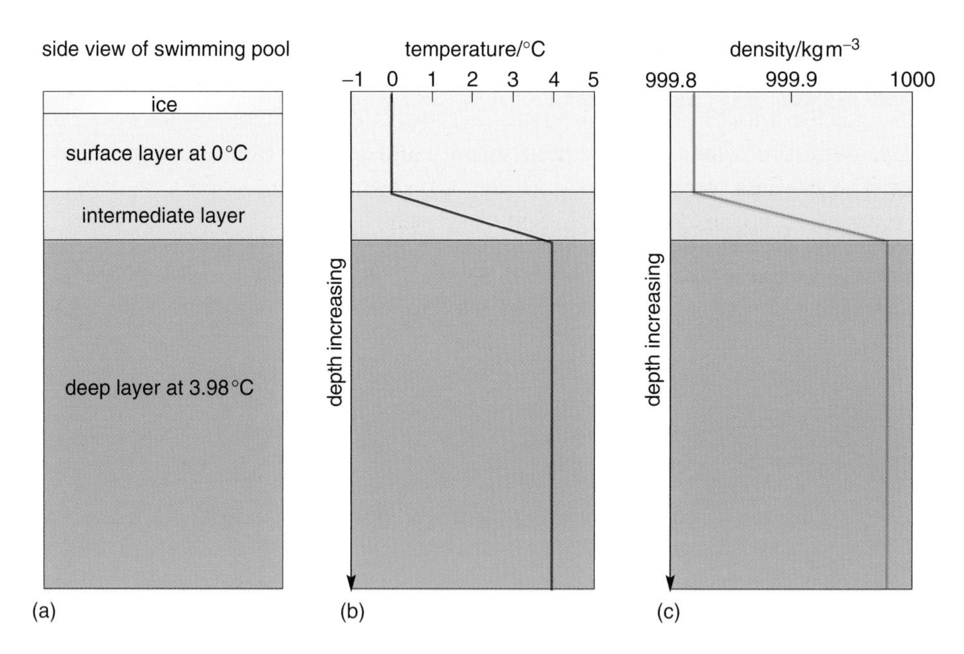

Figure 3.2 (a) Side view of a swimming pool showing ice on the surface. (b) Temperature structure throughout the pool showing the temperature of the three layers: a cold surface layer, an intermediate layer and a deeper warm layer. (c) Density structure mirroring the shape of the temperature profile in (b).

You can see in Figures 3.2b and c that both the temperature and the density increase with depth from the surface to the bottom of the swimming pool. At the bottom of the pool the freshwater is at its maximum density, and the temperature of the water is 3.98 °C.

Let us think about this a little more. Usually when water is heated the water molecules are given more energy and become more widely spaced. If the molecules were more widely spaced in the same volume, the density of the liquid would decrease. But in freshwater the opposite may happen, depending on the temperature. When freshwater at 1 °C is heated to 3.5 °C (Figure 3.1) the density increases. Similarly when a substance is cooled down the molecules usually become more closely packed together and, for the same volume, the density increases. But we know from our swimming pool example (Figure 3.2) that the solid form of water, which is of course ice, is less dense than the water and the ice floats. So below a temperature of 3.98 °C cooling results in the molecules of water becoming more widely spaced and both the liquid and solid forms of water below this temperature expand. This is the reason why water in rock cracks leads to rock shattering and breaking in cold temperatures.

The reason for these anomalous properties is because of the molecular structure of water. A molecule of water is shown in Figure 3.3.

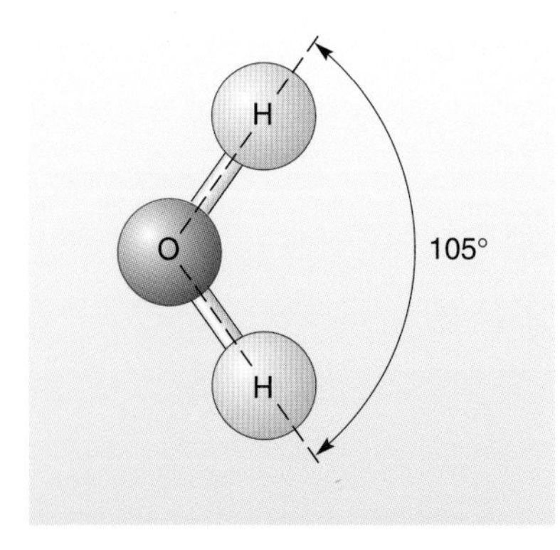

Figure 3.3 A molecule of water, H_2O, with the hydrogen atoms and the oxygen atom sharing electrons. The side of the molecule with the oxygen atom carries a small negative charge and the side containing the hydrogen atoms a small positive charge.

The oxygen atom and the hydrogen atoms share electrons; the angle between the two hydrogen atoms is 105°. This results in a small net negative charge on the oxygen side of the molecule, and a small net positive charge on the hydrogen side of the molecule. This is called a polar structure, and the molecules are weakly attracted to each other. Weak bonds called hydrogen bonds develop. These bonds give a more ordered packing of water molecules at low temperatures. If the temperature is increased (but still below 3.98 °C) the resulting increase in internal energy breaks the weak hydrogen bonds and the H_2O molecules pack together more closely, more fit into the same volume and so the density is increased. Above 3.98 °C the increase in internal energy is such that the molecules really do become more widely spaced and, following Figure 3.1, the density of the water will decrease.

We have already noted that the most obvious difference between seawater and freshwater is that seawater is salty. What is the effect of this salt on the density structure of the water? Figure 3.4 shows three profiles of the temperature, salinity, and density measured in the ocean at latitude 20° S in the Atlantic Ocean. The profiles were taken over a deep abyssal plain almost 5300 m deep; measurements were made to within 5 m of the sea floor.

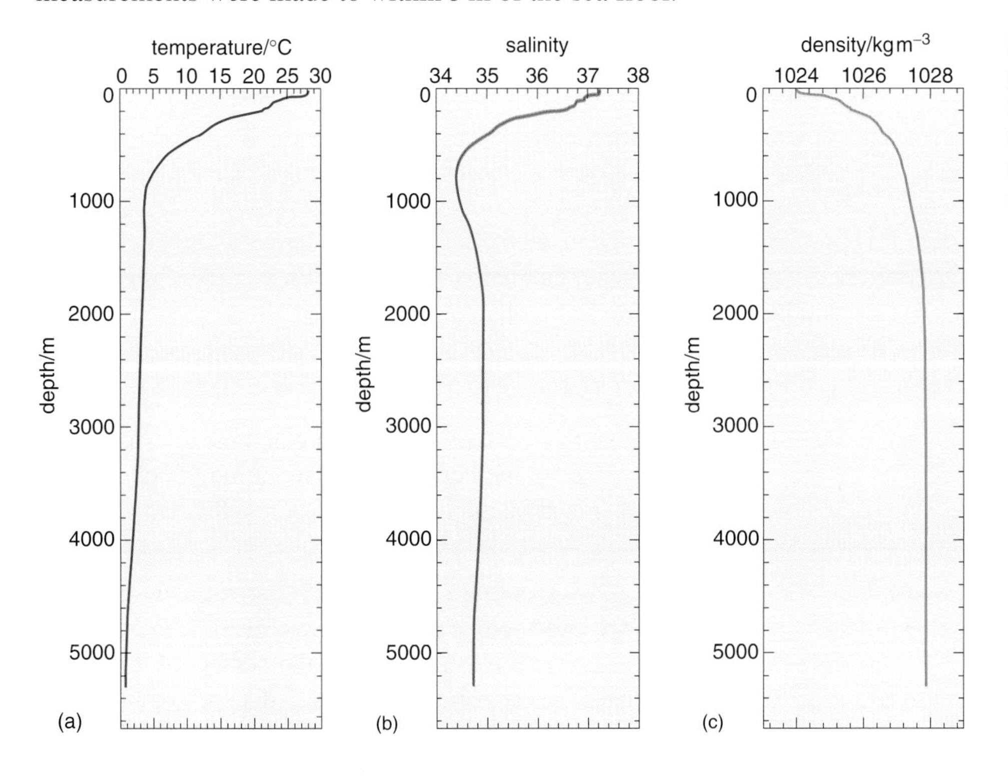

Figure 3.4 Profiles of the physical properties of seawater at 20° S in the Atlantic Ocean: (a) temperature against depth; (b) salinity against depth; (c) density against depth.

- In Figure 3.4, what is the temperature of the seawater at the sea floor?
- The temperature is below 1 °C.

Figure 3.4a shows that the temperature of seawater in this location, unlike that of freshwater, decreases throughout the water column, from almost 30 °C at the surface to below 1 °C at the sea floor. Clearly the addition of salts to the water has changed its physical properties. Figure 3.4b shows that the salinity structure of the water column is very different to the temperature. There is a relatively high surface salinity that decreases to a minimum at approximately 800 m depth. Below this depth the salinity increases to a higher value at approximately 2000 m depth and then it decreases gradually down to the sea floor. In Section 3.7 we shall see how this unusual vertical distribution of salinity arises. Figure 3.4c shows the density of the water column at this location. The density range is from 1024–1028 kg m^{-3}, which is 24–28 kg m^{-3} more dense than freshwater. Density increases in a downward direction from the surface to the sea floor and the water is stratified (see Figure 3.2c). Below 3000 m depth the density does not appear to increase very much at all in Figures 3.2c and 3.4c. However, on a graph that uses a higher resolution you would see that density increases very slightly with depth, and that the whole water column is stratified.

As the density of seawater is typically in the range 1024–1028 kg m^{-3}, we usually use the in-situ density or the density anomaly. This is given the symbol σ_t.

The formula for σ_t is given by:

$$\sigma_t = (\rho - 1000)\ \mathrm{kg\,m^{-3}} \quad (3.2)$$

where ρ is the density of water.

- ❍ What is the range of the density anomaly for typical seawater?
- ● The density anomaly for typical seawater is in the range of 24–28 kg m^{-3}.

The pressure within the ocean is simply proportional to the weight of the water above. This pressure is given by the hydrostatic equation:

$$P = -g\rho z \quad (3.3)$$

where P is pressure, z is a change in depth, and g the acceleration due to gravity. The formula is negative to indicate that the vertical co-ordinate z (depth) is positive in an upward direction.

Box 3.1 How do we measure temperature, salinity and density in the ocean?

Much of our knowledge about how the oceans circulate is based on measuring the various parameters of the water column. The most important instrument is the CTD, which measures the *conductivity* and *temperature* of the seawater, and *pressure* (Figure 3.5). The depth of the instrument is derived by re-arranging Equation 3.3. Other parameters can also be measured, such as sediment particle density and the amount of chlorophyll present (algae).

The CTD is deployed from a ship and lowered on a winch from the surface of the ocean to the sea floor. It is usually lowered quite slowly (about 60 m/minute) in order to get the maximum vertical resolution of data. If the ocean is 5300 m deep as in Figure 3.4, then a round trip (from the surface to the sea floor and back again) would be 10 600 m. So measurement of the vertical profile of temperature, salinity and density in the deep ocean can take almost 3 hours.

Figure 3.5 A CTD instrument. The bottles on the side of the instrument collect samples of water and can be closed by the operator at any chosen depth.

3.4 Temperature in the oceans

The CTD measurement for 20° S shown in Figure 3.4 shows quite a complicated temperature structure. Not surprisingly the only significant source of energy to heat the seawater is from the absorption of solar radiation arriving at the surface (Figure 3.6).

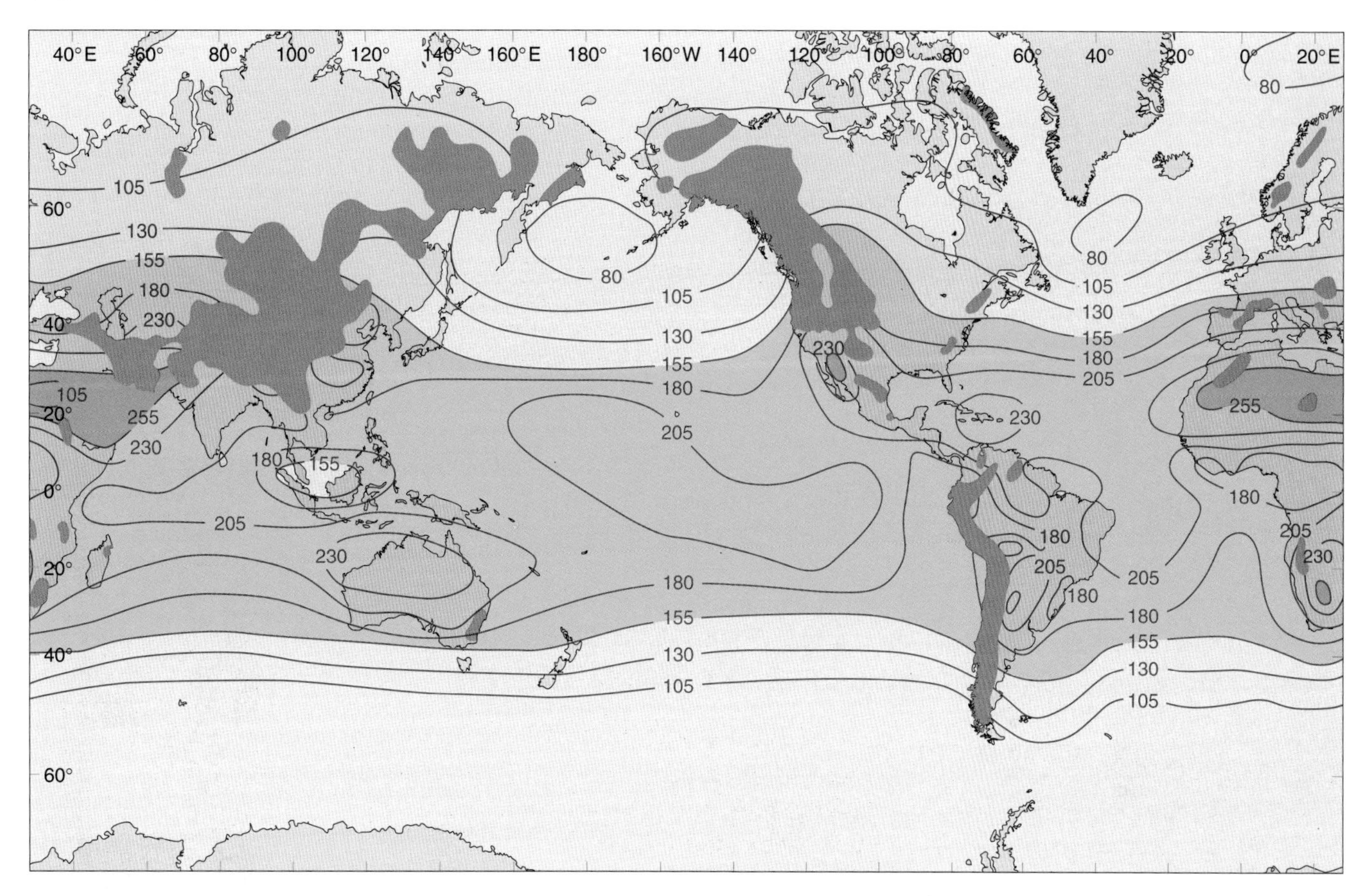

Figure 3.6 The amount of solar radiation that reaches the surface of the Earth ($W\ m^{-2}$) averaged over one year. Contours from high mountain regions have been omitted.

As you saw in Block 2, Part 1, the albedo of water is quite low, so most of the solar radiation received at the surface of the oceans is absorbed and heats the surface seawater. This sea-surface temperature (SST) can be measured from space by satellite (Figure 3.7).

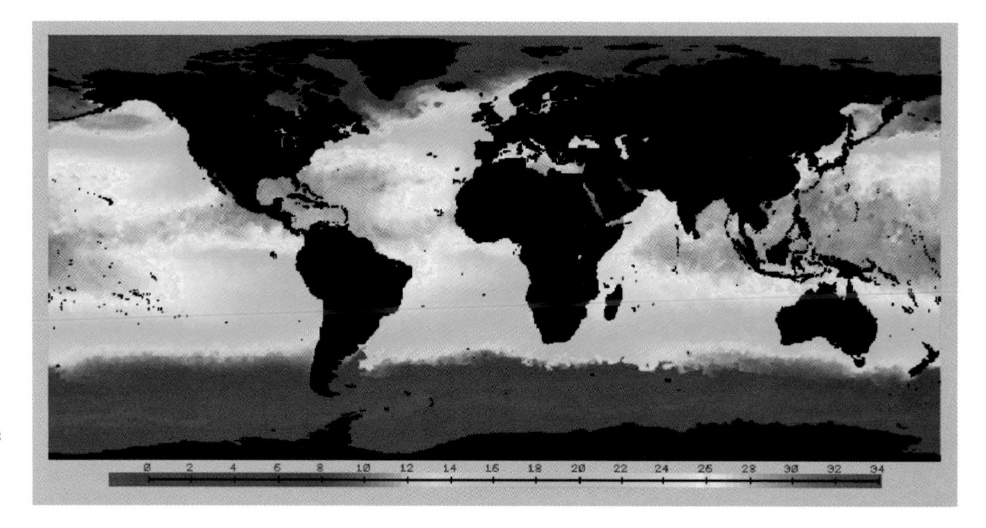

Figure 3.7 A map of the sea-surface temperature of the Earth from the satellite NOAA–16 for the period 1–4 September 2001. The scale is in °C.

In Figure 3.7 the sea-surface temperature has a false colour scale, but areas with the same colour have the same temperatures. In general terms the SST is colder at high-latitudes, and warmer in mid-latitude and equatorial regions. By comparing Figures 3.6 and 3.7, how well do you think the incoming solar radiation in Figure 3.6 relates to the shape of the colours showing equal temperature in Figure 3.7? You might expect those regions with high incoming solar radiation to have the highest SST. This is generally true. However, if you look more closely you can see that there are in fact large departures between the two plots. Take a look at the coast of Portugal in Figures 3.6 and 3.7. In Figure 3.6 the average incoming annual solar radiation is 180–205 $W\,m^{-2}$, yet in Figure 3.7 the SST is 18–20 °C. If you now look at the southwest coast of Africa in Figure 3.6 you can see that the average incoming annual solar radiation is again 180–205 $W\,m^{-2}$, yet in Figure 3.7 you can see that the SST is 13–15 °C, which is much colder than the SST off the coast of Portugal.

Question 3.2

Can you think of any reasons why the SST field in Figure 3.7 does not match up with the average incoming annual solar radiation in Figure 3.6?

The contours of Figure 3.6 are almost all along lines of equal latitude whilst in Figure 3.7 the colours indicating equal temperatures follow a different pattern. The main reasons for this will be discussed in Section 4.

3.5 The vertical distribution of temperature with depth

The surface of the ocean is warmed by the Sun. We have already seen (Figure 3.4) that there is a variation of temperature with depth.

❍ How can the energy from solar radiation be carried from the surface to deeper in the ocean?

● The energy can be transported downwards by conduction and by wind mixing.

The amount of conduction in the ocean is both small and a slow process, so it can effectively be ignored. However, the effect of the wind on the surface of the ocean is very significant. Wind creates turbulence at the surface of the ocean and the warmer water is mixed up to create a layer of almost constant temperature called the **wind-mixed layer**. Figure 3.8 shows an expanded plot of the top 200 m of the CTD profile shown in Figure 3.4.

Figure 3.8 shows that at 20° S the mixed layer at the surface of the ocean is approximately 35 m thick and its temperature is slightly more than 28 °C. In some parts of the world where the winds are particularly strong, the mixed layer can be up to 200 m thick. Beneath this mixed layer is a region where the temperature decreases from 28 °C to approximately 5 °C at a depth of about 1000 m (Figure 3.4). This region of rapidly decreasing temperature is called the **permanent thermocline**. Beneath the permanent thermocline the water temperature decreases to between 0 °C and 3 °C at the sea floor. We have looked at a temperature profile in one location, but what is the temperature distribution like across the whole Atlantic Ocean? Such a section is shown in Figure 3.9.

Figure 3.9 shows a **hydrographic section** across the ocean. This is the standard way for presenting a series of CTD measurements taken in a straight line. This section actually includes data from over 120 CTD stations, including the station shown in Figure 3.4. The section is created by plotting the latitude along the *x*-axis — and in Figure 3.9 this is over 11 000 km long. The *y*-axis represents the depth. Colours are used to represent equal temperatures measured at the CTD stations. The black lines of equal temperature — called isotherms — are superimposed on the colours. The isotherms at 0 °C, 10 °C and 20 °C are identified by bold lines and white numbers. The spiky grey shaded region at the bottom of the plot represents the sea floor along the section. The red line on the inset map of the Atlantic Ocean shows the location of the hydrographic section.

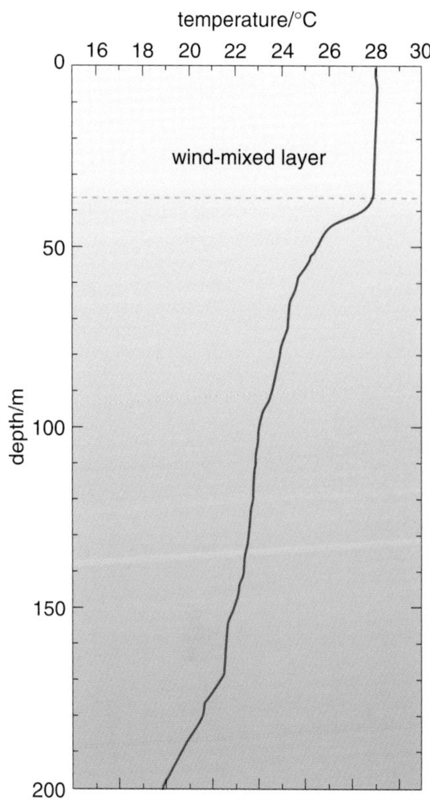

Figure 3.8 The top 200 m of the temperature data recorded as part of the CTD profile at 20° S in the Atlantic Ocean.

The hydrographic section clearly shows that the range of temperatures in the upper 1000 m of the water is much greater than the range of temperatures below 1000 m down to the sea floor. The temperature can change as much as 20 °C in the upper 1000 m. Above 1000 m the isotherms are 'W' shaped, that is they are generally shallow at high latitudes above 40° N and 40° S, are at their deepest points at about 30° N and 30° S, and are shallow again at the Equator. We shall see in Section 4 that this 'W' shape is a feature of the wind field across the Atlantic Ocean.

From 1000 m to the sea floor the temperature change is much smaller and in the range 4–5 °C. The isotherms here have a 'U' shape across the Atlantic Ocean. At latitudes higher than 40° S and 40° N the isotherms bend up towards the surface. The isotherms that are in the vertical range of 1000–2500 m are at their deepest point at the Equator. The mauve-coloured region in the south that is close to the surface at 50° S is more of an 'L' shape and is very cold water from the Antarctic that is descending rapidly and spreading throughout the deep ocean. This mauve-coloured region is clearly stopped by the seamounts of the Mid-Atlantic Ridge. This ridge can be seen at the Equator and

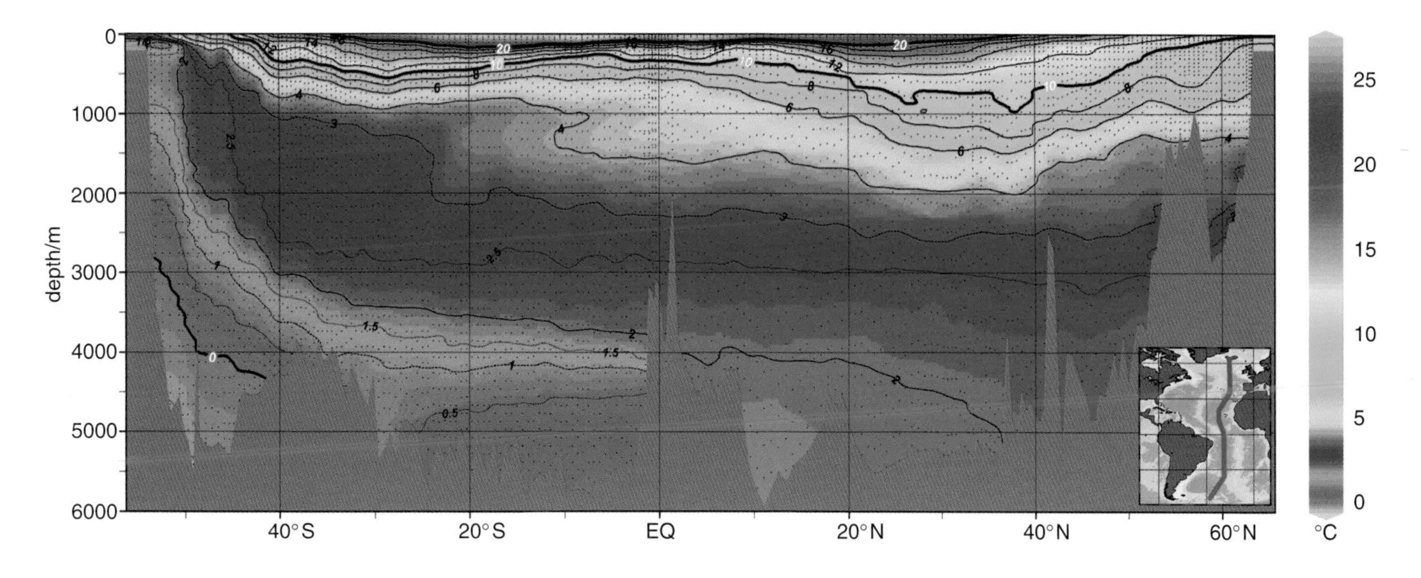

Figure 3.9 A south–north temperature section across the Atlantic Ocean from 55° S to 65° N. The CTD profile shown in Figures 3.4 and 3.8 is included in this section at 20° S.

on the inset map of Figure 3.9. In this mauve region of very cold water, you can see the 0 °C isotherm from approximately 2900 m at 53° S descending to 4400 m at 42° S. Water beneath this isotherm is below 0 °C.

- Does this mean that there is ice in the ocean at this depth?

- No, there is no ice at this depth. Just as in freshwater, the ice would float because of the molecular structure of H_2O. What has happened is that the addition of salts has lowered the freezing point of the water to below 0 °C.

This very cold water is formed at the surface of the ocean in both the Arctic and the Antarctic, and sinks to the ocean depths. In the Antarctic this cold water spreads out throughout the Atlantic Ocean and is easily seen in Figure 3.9.

Question 3.3

Can you think of any reasons why very cold water formed in the Arctic does not spread out through the deep ocean? (Hint: look at your poster map 'The Earth's surface'.)

Finally, we can see in Figure 3.9 that at the Equator the temperature between the surface and the sea floor varies over a wide range — almost 30 °C in places. As we go to higher latitudes the range in temperature between the surface and the sea floor is very much reduced. In Figure 3.9, at about 52° S the range is only approximately 3.5 °C, at 60° N the range is larger at about 10 °C, but north of the ridge system between Greenland and Scotland (which holds back the cold water) the vertical range of temperature is also about 3.5 °C.

We have seen that solar radiation is mostly responsible for heating the surface of the ocean, and the lack of solar radiation in the high-latitude regions is responsible for cooling the surface. Waters of different temperature are then spread throughout the oceans because colder water is more dense. Cold water sinks beneath the warmer waters to create the picture seen in Figure 3.9. Could there be a similar process that is determining the salinity at the surface of the oceans?

3.6 Salinity in the oceans

Just as temperature can vary with depth, the distribution of salinity with depth also varies (Figure 3.4). The only significant modification of salinity takes place at the surface of the ocean. There are two competing processes that determine surface salinity: *precipitation* and *evaporation*. By precipitation we mean any process that can reduce the salinity of the surface waters, such as rain on the surface of the ocean and rivers flowing into the seas.

If we work out the amount of water added to the global ocean by precipitation, P, through rain and rivers, as a value in mm, and work out the amount of water being removed by the process of evaporation, E, also as a value in mm, we can calculate the value of $E - P$. This value determines the salinity of the surface waters. The values of $E - P$ and the resulting approximate global-surface salinity distribution is shown in Figure 3.10.

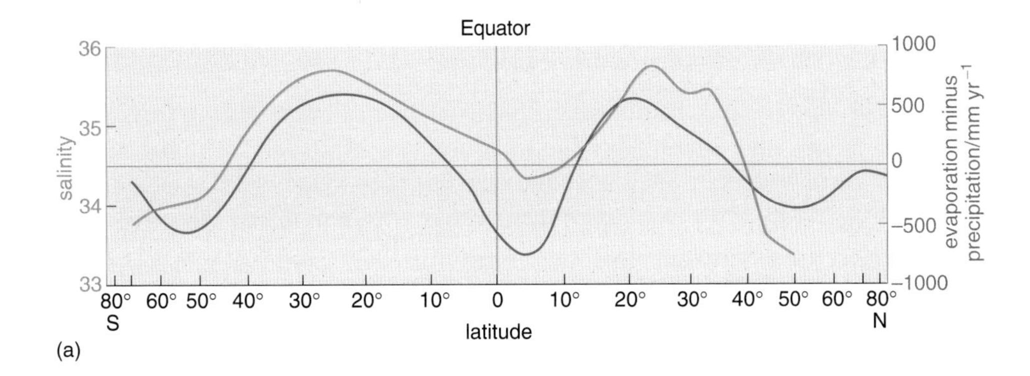

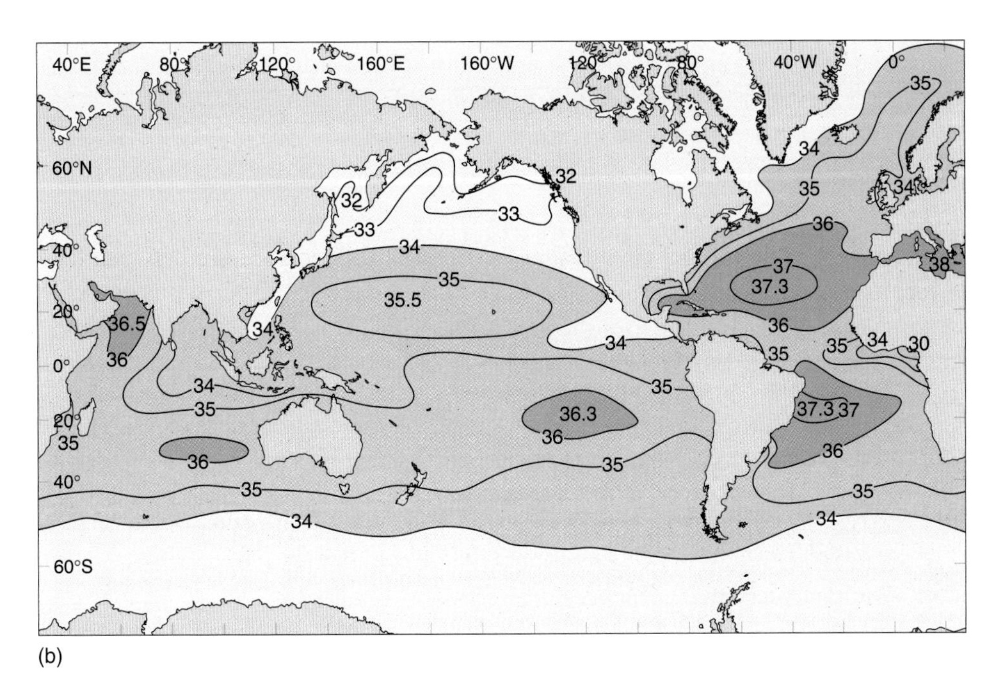

Figure 3.10 (a) The mean value of $E-P$ against latitude (purple line) and typical values of the surface salinity (green line). (b) Mean annual values of the sea-surface salinity.

The mean value of $E-P$ against latitude is shown in Figure 3.10a. If the value of $E-P$ is negative then the sea-surface is freshened; when $E-P$ is positive the salinity of the surface waters is increased. The shape of the purple line is perhaps not what you would be expect. Typically air temperatures are much greater at low latitudes (as can be seen from the incident solar radiation shown in Figure 3.6) and so we would expect $E-P$ to be positive, that is evaporation would be greater than precipitation, but this is clearly not the case. Tropical rainfall freshens the surface water and in the equatorial Atlantic the outflow of the river Amazon has an additional strong influence. Both of these effects have changed the shape of the curve. The values of $E-P$ has an 'M' shape between the latitudes of 60° N and 60° S, the trough being due to tropical rainfall and river inputs. The two peaks of positive $E-P$ values, at about 25° S and 20° N, are due to the decrease in rainfall and high evaporation at these latitudes. The $E-P$ value then falls at mid-latitudes where rainfall increases. At latitudes higher than about 60° the decrease in rainfall, coupled to a decrease in evaporation caused by falling temperatures, controls the shape of the curve. We can see in Figure 3.10a that the $E-P$ curve does not match exactly the surface salinity curve. Clearly, as in the case of the sea-surface temperature, there is another factor influencing the surface salinity. This is the circulation of the oceans.

Figure 3.10b shows a map of the mean annual surface salinity values across the world. We do not have instruments that measure the surface salinity in the same way that we can measure SST, and so the map has been made in the same way that Matthew Maury drew his map of the oceans — by joining up regions with equal values, which in this case are salinity values. Figure 3.10b also shows a striking difference between the Atlantic and Pacific Oceans: the salinity in the Atlantic is much higher. In the Pacific Ocean, lines of equal salinity, which are called isohalines, tend to follow lines of latitude whereas in the Atlantic there is a clear departure. In both cases it is a feature of the circulation of the ocean. You can also see that off the coast of Portugal there is evidence of ocean circulation as less saline waters from the northwest are swept southwards. At very high latitudes the surface salinity decreases rapidly as the input from rivers is high. There is also an additional input of freshwater from the melting of ice.

3.7 The vertical distribution of salinity with depth

The salinity profile also has a wind-mixed layer, which is shown in Figure 3.11.

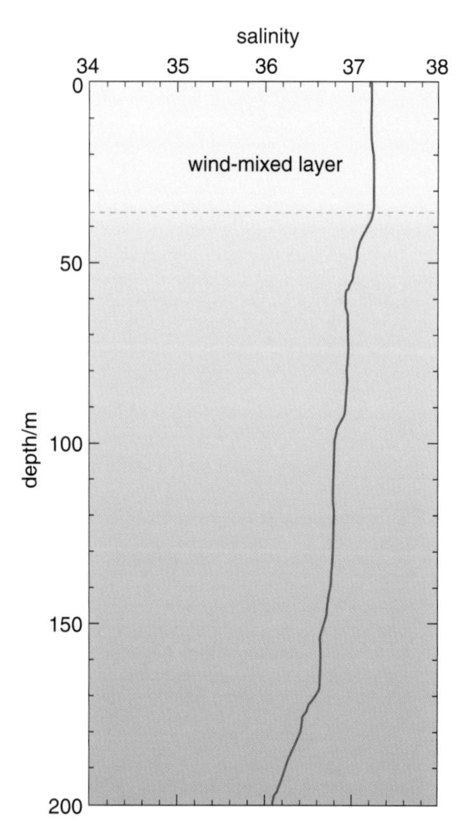

Figure 3.11 The top 200 m of the salinity data from the CTD profile recorded at 20° S in the Atlantic Ocean.

The depth of this layer coincides with the depth of the mixed layer seen in the temperature profile (Figure 3.8). Beneath this mixed layer salt can diffuse very slowly through the water column, or be moved about by ocean currents. These ocean currents are responsible for the shape of the salinity profile from the surface to the sea floor seen in Figure 3.4. Figure 3.12 shows a salinity section across the Atlantic Ocean that was recorded simultaneously to the temperature section in Figure 3.9.

The salinity distribution across the Atlantic Ocean shows a much more complicated structure than that shown by the temperature section. Instead of being able to describe the shape of the isohalines in terms of letters of the alphabet (like we did for temperature), it is almost as if there is a giant clockwise circulation across the whole of the Atlantic Ocean that is wrapping the colours up. The very low-salinity, mauve-coloured water at the surface at about 50° S in Figure 3.12 sinks down to a region centred on approximately 900 m, and has been dragged north of the Equator by the circulation pattern. The low-salinity water at about 800 m depth in the salinity profile in Figure 3.4 appears to have come from the Antarctic! As the water appears to move to the north the colours change from mauve to blue, which indicates that the salinity is increasing. North of this blue 'tongue' of water, in the region 20–40° N, the orange–yellow water of salinity about 35.5 seems to be dragged down in this clockwise circulation. Beneath this higher-salinity water in the region 3000–4000 m depth is less saline water that appears to be dragged to the south. As the water appears to move to the south, the depth of this patch of water is decreasing.

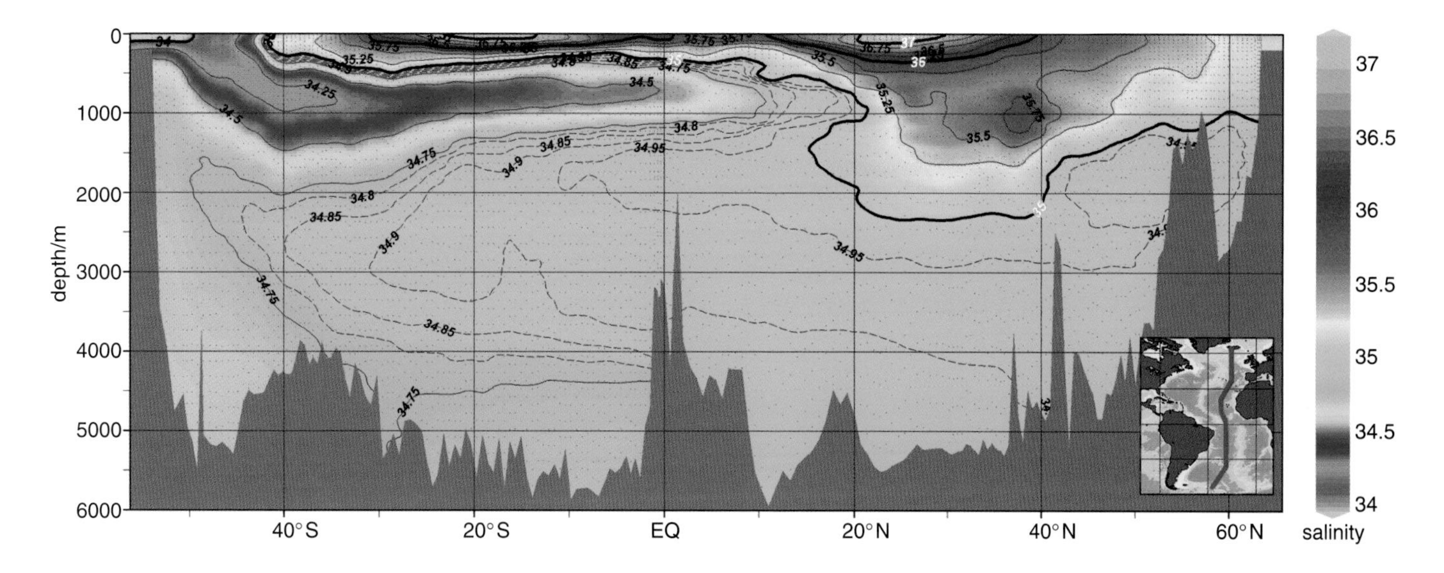

Figure 3.12 A south–north salinity section across the Atlantic Ocean from 55° S to 65° N. The CTD profile shown in Figures 3.4 and 3.11 is included in this section at 20° S.

The water that is filling up the South Atlantic below 4000 m must be more dense than the fresher water that appears to come from the north. In Section 4 we shall see how this pattern of apparent circulation arises. Finally it is worth noting once again, that although Figure 3.12 shows that the salinity across an ocean basin can have a range of 34 to 37, if you took a sample of water from several places and analysed them for the ionic constituents shown in Table 3.1, the ratios of the amount of each salt listed in Table 3.1 would be constant.

3.8 Density in seawater

You may have guessed already that density is a function of both temperature and salinity. The temperature and salinity data in Figure 3.4 have been used to calculate the density profile that is shown in Figure 3.13. This Figure shows the top 200 m of the density profile shown in Figure 3.4.

Figure 3.13 shows that the almost constant temperature and salinity in the top 35 m of the ocean have created a layer of water that is of almost uniform density, in this case approximately 1024.1 kg m^{-3}. This is the wind-mixed layer. Beneath this, the decrease in temperature, and associated salinity, increases the density rapidly at first, and then more slowly. This layer, where density increases rapidly, is called the pycnocline. The relationship between temperature and density for seawater is very complex and non-linear. Figure 3.14 shows a plot of density against temperature for seawater of different salinities.

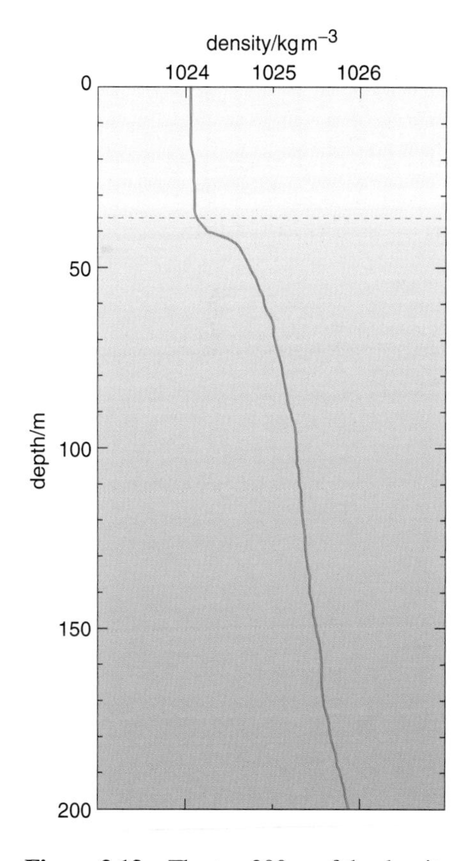

Figure 3.13 The top 200 m of the density profile derived from the CTD profile recorded at 20° S in the Atlantic Ocean.

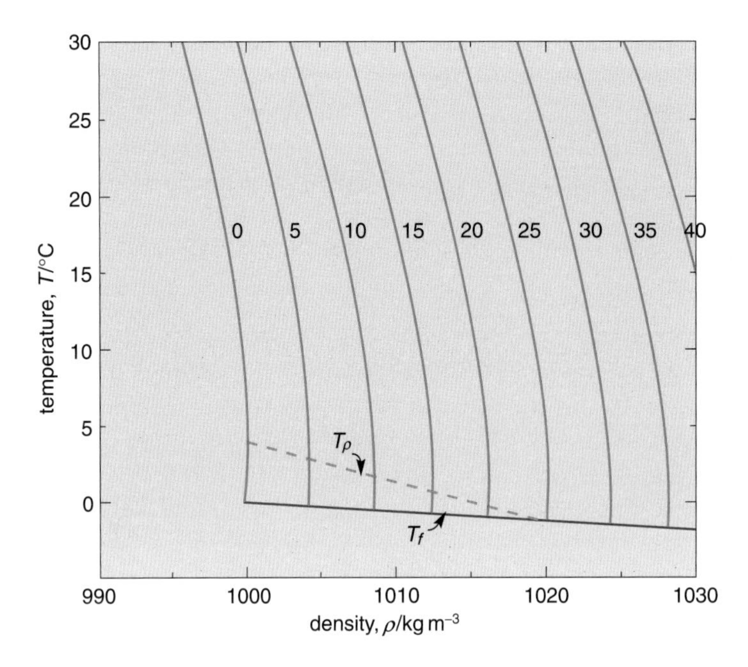

Figure 3.14 The relationship between density and temperature for different values of salinity. Also shown are the temperatures at which seawater freezes, T_f, and the temperatures at which density is a maximum, T_ρ.

Figure 3.14 shows a family of curves. There is one curve of density values against temperature for each value of salinity. The numbers 0, 5, 10, etc. are salinity values. If you look at the first curve on the left-hand side of the plot you will see it is labelled 0. This means the salinity is zero and the curve represents freshwater. This first curve is the same as Figure 3.1 except that the axes have been exchanged. In Figure 3.14 you can see that when the salinity value is 0, the maximum density of the water is, as you would expect, at 3.98 °C. A straight dashed line, labelled T_ρ, joins the density curve at this temperature. This line represents the temperatures of maximum density. The first curve is brought to a stop at a temperature of 0 °C because this is, as you know, the temperature at which ice is formed in freshwater. To represent this, a solid red line, labelled T_f, touches the density curve at this point. If you now look at the second curve, where the salinity value is 5, you can see that the temperature at which the density is a maximum, T_ρ, has decreased below 3.98 °C, and the freezing temperature, T_f, has dropped to below 0 °C. You can also see that close to the freezing temperature the gradient of the density curve is different for a salinity value of 0 when compared to 5. This is a demonstration of the non-linearity and, although density is a function of temperature and salinity, there is no simple equation to show their relationship.

As the salinity increases the temperature of maximum density, T_ρ, decreases until a salinity value of 24.7 is reached. At this point the temperature of maximum density, T_ρ, is equal to the freezing temperature, T_f. If you think back to the example that we described in Figure 3.2, when the salinity value was 0, you may remember we ended up with a bottom layer of dense water in the swimming pool of 3.98 °C. If the water in the swimming pool had a salinity value of 5, the same process that we discussed above would take place and ice would form from a less dense surface layer. The difference is that this time the temperature of the water at the bottom of the pool would be less than 3.98 °C. From Figure 3.14 we can see that this temperature would be around 3 °C. As the salinity increases the temperature of the dense water at the bottom beneath the ice decreases until a salinity of 24.7 is reached. At this point the temperature of maximum density, T_ρ, is equal to the freezing temperature, T_f, and we no longer have to cool all the water in the pool to the temperature of maximum density before ice will form, and there will be no warm dense layer of water at the bottom of the pool. When the salinity is higher than 24.7, the water becomes more dense as it cools, but if the water beneath is more salty, it will be more dense and so the surface water cannot sink (see Figure 3.14).

Compared to freshwater bodies such as large lakes, ice will form more easily on the ocean. This has a special importance in the circulation of the oceans because when ice is formed on the surface of the water in polar regions, the salt is squeezed out of the ice crystals. This salt then increases the density of the surface waters. This more dense seawater will sink from the surface and flow into the interior of the ocean to create circulation patterns like those in Figure 3.12.

Question 3.4

Using Figure 3.12, shade in the region of densities on Figure 3.14 that you would expect to find in the oceans.

3.9 Summary of Section 3

1 Seawater contains dissolved salts made up of a mixture of many different ionic constituents. The ratio of each ion is constant whatever the salinity. This is called constancy of composition.

2 The amount of salt in the ocean has been approximately constant for 10^8 years. Erosion of rocks and volcanic gases both contribute salts to the ocean.

3 Water is a liquid with very unusual physical and chemical properties. Because the H_2O molecule has a polar structure, the maximum density is *not* at the freezing point. This means that in freshwater the entire water column must be cooled before ice can form.

4 Seawater is heated at the surface by solar radiation, although maps of the sea-surface temperature do not necessarily reflect maps of incoming solar radiation. The vertical distribution of temperature across the Atlantic Ocean shows that the isotherms have a 'U' shaped structure, with colder waters at high latitudes.

5 The salinity of seawater is controlled by the balance between evaporation and precipitation, $E - P$. Evaporation increases salinity and precipitation (including river input) reduces the salinity.

6 The vertical distribution of salinity across the Atlantic Ocean shows a complicated structure that implies a large vertical circulation pattern across the whole ocean.

7 Density in the ocean is a non-linear function of temperature and salinity. In the mixed layer there is uniform temperature, salinity and density. Beneath this mixed layer the density increases all the way to the sea floor and the waters are said to be stratified.

Learning outcomes for Section 3

After working through this section you should be able to:

3.1 List the sources of the different ionic constituents that make up the salts in seawater.

3.2 Describe what is meant by the principle of constancy of composition, and be able to explain why we can measure just one ionic constituent to determine how much salt is in seawater.

3.3 Describe how we know that the salinity in the oceans has not changed for the last 10^8 years.

3.4 Explain why the density of freshwater decreases below a temperature of 3.98 °C.

3.5 Explain the reasons why a map of sea-surface temperature would not match up with the incoming solar radiation.

3.6 Describe the distribution of temperature with depth across the Atlantic Ocean and be able to explain why cooler water is at the sea floor.

3.7 Explain how the balance of $E - P$ controls surface salinity of seawater.

3.8 Explain why the whole water column in the ocean needs to be cooled before the ocean freezes.

3.9 Describe why you would expect there to be a layer of uniform temperature, salinity and density in the surface water of the ocean.

Ocean currents

4

In Section 1 you looked at an example of the effect of ocean currents on regional climate. In Section 3 we described how, as well as transporting heat, ocean currents can affect the surface distribution of both temperature (Figure 3.7) and salinity (Figure 3.10). In this section we look at the processes that set up the surface currents and the critical importance of the shape of the ocean basins in driving the three-dimensional, global ocean circulation.

4.1 How do we know the water moves?

Mysterious and unusual items washed up on the beaches of the world are an indication that ocean currents move the waters of the Earth. Tree trunks, for example, are often washed up on the beaches of eastern Greenland — yet no trees have grown in this region for hundreds of years.

The effect of the ocean currents became clear with the first sailings across the seas. Early sailors often ended up at a different location from the one to which they intended to sail. This was because they had no knowledge of the direction of the ocean currents. If you expect to land at a wide and safe bay then landing at a rocky headland would be quite a surprise! The sailors knew there was another force other than the wind that must be pushing the ship — the ocean must be moving (Figure 4.1).

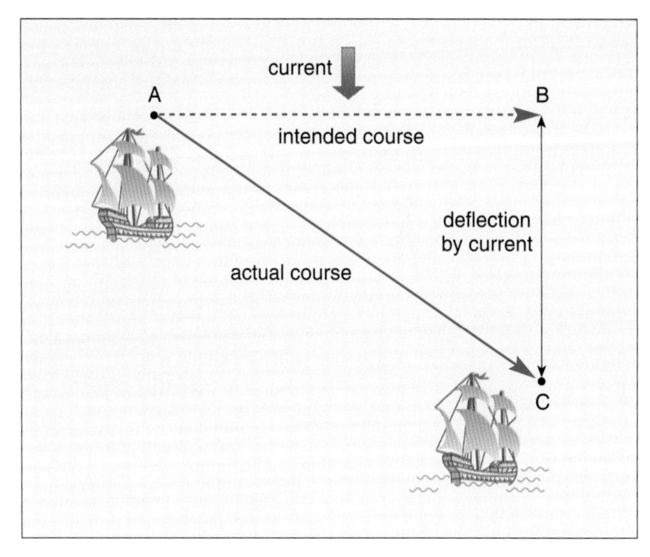

Figure 4.1 The drift of a ship caused by an ocean current. When there is no ocean current there is no deviation from the intended route and the ship sails from A to B. When there is a current the ship deviates and, although pointing in the direction AB throughout, is displaced by the ocean current and ends up at location C. The displacement is given by the length of the line BC.

If the ship in our example is pointing in the direction AB throughout, when there is no ocean current the ship will arrive at its intended location, which is point B. However when there is an ocean current the ship will be deflected from its intended path and will end up at location C. The ship has been deflected from its intended path by the effect of the ocean current. The displacement of the ship from its intended destination, that is the length of the line BC, provides information about the strength of the ocean current.

When sailors reached land, once they recognized where they had arrived, it was straightforward to work out how much the actual course had deviated from the intended course, and therefore easy to work out the strength of the current that had caused the deviation BC. To work out the distance BC is much harder out of sight of the land as it critically depends on the sailors knowing the exact locations of points A, B and C.

The method of navigating with only the course steered and the distance travelled known is called **dead reckoning**. In the case of Figure 4.1, if the ship were travelling at $10\,\text{km}\,\text{h}^{-1}$ for 1 hour, the distance travelled from A to B would be calculated as follows:

$$\begin{aligned}\text{distance travelled} &= \text{speed} \times \text{time} \qquad (4.1)\\ &= 10\,\text{km}\,\text{h}^{-1} \times 1\,\text{h}\\ &= 10\,\text{km}.\end{aligned}$$

This means that the location of the ship after 1 hour, that is the dead-reckoning position, would be at point B — which is exactly 10 km to the east of point A. If the ship deviated from its intended course due to an ocean current and ended up at point C, it is not possible to work out the deviation (that is the length of the line BC) unless the location of point C is accurately known. This is the *limiting factor*.

❍ With reference to Figure 4.1, if you can determine your location to an accuracy of 3 km, suggest an easy way to improve your estimate of the ocean current.

● If you could work out your location to an accuracy of 3 km, a simple way to improve your estimate of the ocean current would be to increase the length of time that you travelled before trying to work out the length of the line BC. This would mean that the 3 km accuracy would then be less important as the length of the line BC would be much longer. For example if you travelled for 1 hour at 10 km h^{-1} the length of the line AB would be 10 km, but as we know the location is only accurate to within 3 km, it could be anywhere between 8.5 km and 11.5 km. The error in our calculated position

$$= \frac{3}{10} \times 100\%$$
$$= 30\%$$

If the ship travelled for 100 hours, the length of the line AB

$$= 10\,\text{km}\,\text{h}^{-1} \times 100\,\text{h}$$

$$= 1000\,\text{km}.$$

The relative error

$$= \frac{3}{1000} \times 100\%$$
$$= 0.3\%$$

You would therefore be able to estimate the average ocean current much better because the relative error is much smaller.

There is a trade off with this method of improving the accuracy. The longer the time period used between comparing your dead-reckoning estimate of the ship's location (point B) with the location that you determine for point C, the less we learn about the ocean current that has deflected the ship. This is because the deflection of the ship (BC) will be made up of the average ocean current over the whole time period, and so the resolution of the ocean current deviation is reduced. Nevertheless, this method for deriving mean ocean currents allowed sailors to begin to draw maps of the global ocean's surface movements.

Box 4.1 How is position found at sea?

Once sailors leave sight of land, it is difficult to measure precisely their location on the Earth. It is possible to navigate using the relative location of the stars, the Sun and the Moon — but of course this method requires a clear sky. For this reason the method of dead reckoning was developed (Figure 4.1) and for many centuries was the only way of attempting to determine position.

As methods developed, by the 15th century it became relatively easy for a skilled sailor to work out the distance (the latitude) of a ship from the Equator. This was done by measuring precisely the angle that the sun makes with the horizon when it is at its highest at midday. It is however much harder to measure longitude. In fact it is so difficult that in 1714 the British Parliament passed an Act, which became known as the Longitude Act, setting a prize of £20 000 (equivalent to approximately £2.5 million today) for a 'practical and useful' method of determining longitude.

The method that was ultimately successful, and has endured, is based on the fact that when the Sun is at its highest, that is directly overhead, it is exactly midday at that location. The Earth spins on its axis once in 24 hours. This means that in every hour the Earth rotates (relative to the sun) 360°/24 hours, which is $15° \text{ h}^{-1}$. If the time of local midday is compared to a reference point, such as Greenwich, London:

- for every 1 hour that local midday is *later* than that at Greenwich, the location will be at a longitude of +15° around the globe — in our frame of reference we call this 15° W;
- for every hour local midday is *before* that at Greenwich, the longitude will be −15° from the reference point, we call this 15° E.

This means that the critical step in determining longitude is to measure the time of local midday and compare it to a reference location.

Although not difficult today, the manufacture of a clock that could run without losing time whilst in a moving ship was an incredibly difficult thing to do. The Englishman John Harrison (1693–1776), a self taught mathematician and clockmaker, took up the challenge and made it his life's work. Over a period of almost 25 years Harrison built four clocks that were numbered H-1 (Harrison 1) to H-4 to try and win this fortune. The first was a giant (H-1, Figure 4.2a). It weighs over 30 kg and stands in a box with sides over 1.2 m in length. The clock that actually fulfilled the terms of the Longitude Act was called H-4 (Figure 4.2b) and is as different to H-1 as could be imagined.

H-4 is about 12 cm in diameter and weighs only 1.4 kg. Despite fulfilling the terms of the Longitude Act, the British establishment repeatedly changed the conditions for claiming the prize money, and the prize itself was never formally awarded. However, after the intervention of King George III, John Harrison was eventually paid. His clocks can still be seen today in the National Maritime Museum, Greenwich.

In the 21st century the standard method of ocean navigation is to use a network of more than 30 satellites, called the Global Positioning System (GPS). A receiver that can process these satellite signals can be purchased in almost any outdoor pursuits shop for less than £100. These receivers routinely give accuracy to within 15 m. Perhaps surprisingly, the fundamental principle of the GPS system is also based on the very accurate measurement of time.

(a)

(b)

Figure 4.2 Harrison's two most famous clocks and the dates they were delivered to the British Government. (a) Harrison 1 (H-1), 1737, (b) Harrison 4 (H-4), 1760.

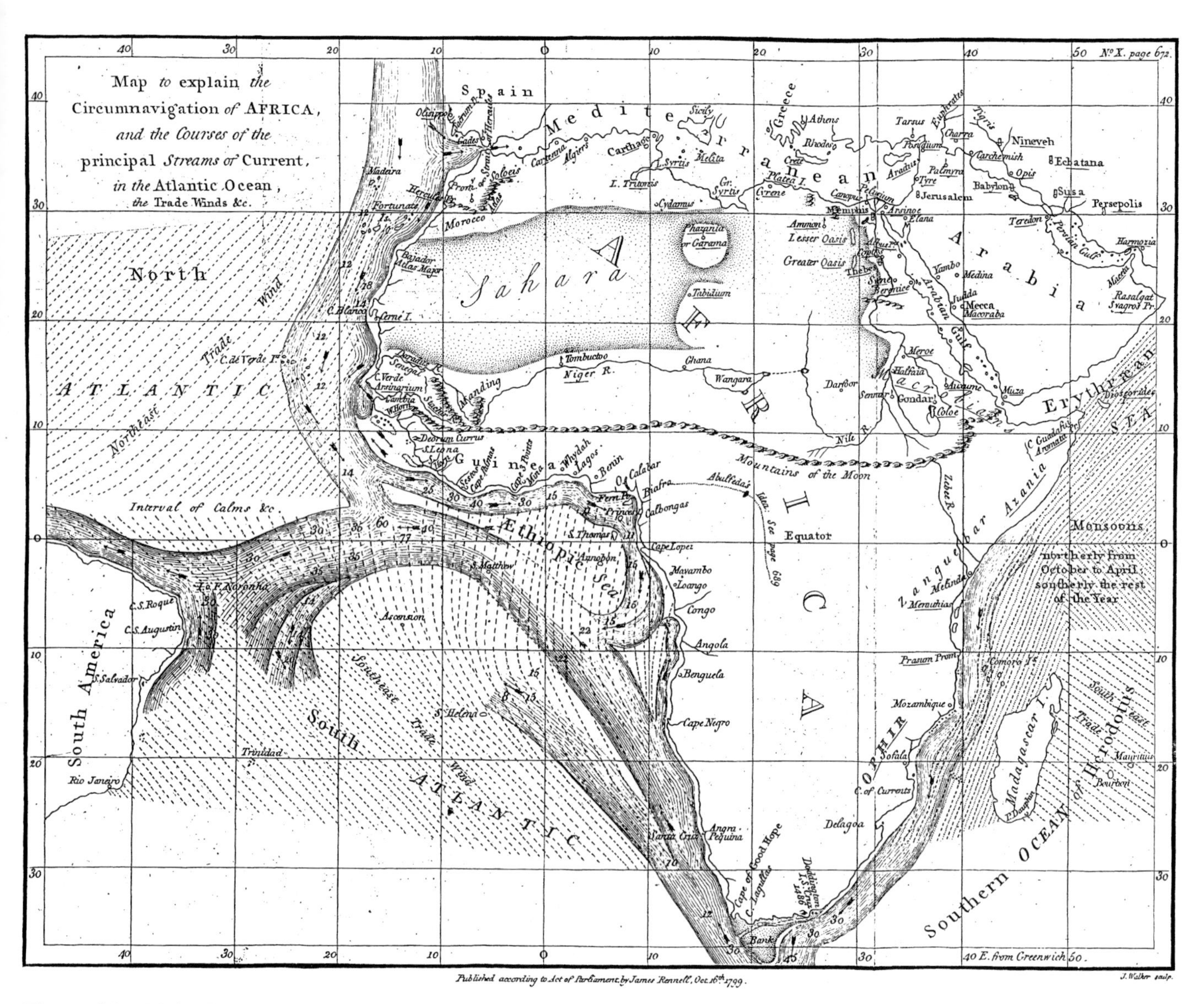

Figure 4.3 Major James Rennell's map of the currents around Africa published in 1800. Dashed lines represent prevailing winds.

The first systematic collection of ocean-current data made by sea captains appeared in the 17th century. One of the most famous was compiled by the Englishman Major James Rennell. Rennell studied the logs of ships that had travelled from the East Indies to England on trade routes and used this data to draw a map of the surface currents around the coasts of Africa.

You can see in Figure 4.3 that Rennell has drawn the stronger ocean currents as being like rivers within the ocean. Solid arrows show the direction of flow; dashed lines represent the prevailing wind directions. In some places, such as the east coast of Africa, the prevailing wind directions appears to match up well with the direction of the surface currents. In other areas, such as The Gulf of Guinea — labelled Ethiopic Sea — and the equatorial Atlantic Ocean, there are clear differences. It is also worth noting the rather unusual geography within North Africa on this chart, for example the River Nile and River Niger are almost joined, and there is a continuous mountain chain across the continent called the 'Mountains of the Moon'. Within 20 years of Rennell drawing this chart the Royal Navy sent many of its sailors into the North African desert to investigate these geographical 'truths'!

What is true is that maps such as the one shown in Figure 4.3 have proved a major asset to ocean navigators by reducing the length of time taken for a ship to travel between destinations. A ship with such a map can choose a route to 'ride' the favourable ocean currents and so greatly increase its speed. The American Matthew Maury drew a similar chart to Figure 4.3, which showed the currents around the Americas. Clipper ships used it to reduce the length of time taken to get from New York to San Francisco from an average of 180 days to just over 130 days. With such a huge saving in time the commercial importance of oceanic-current maps is obvious.

Oceanographers have managed to create modern maps of the surface currents by putting together many years of oceanographic research (Figure 4.4).

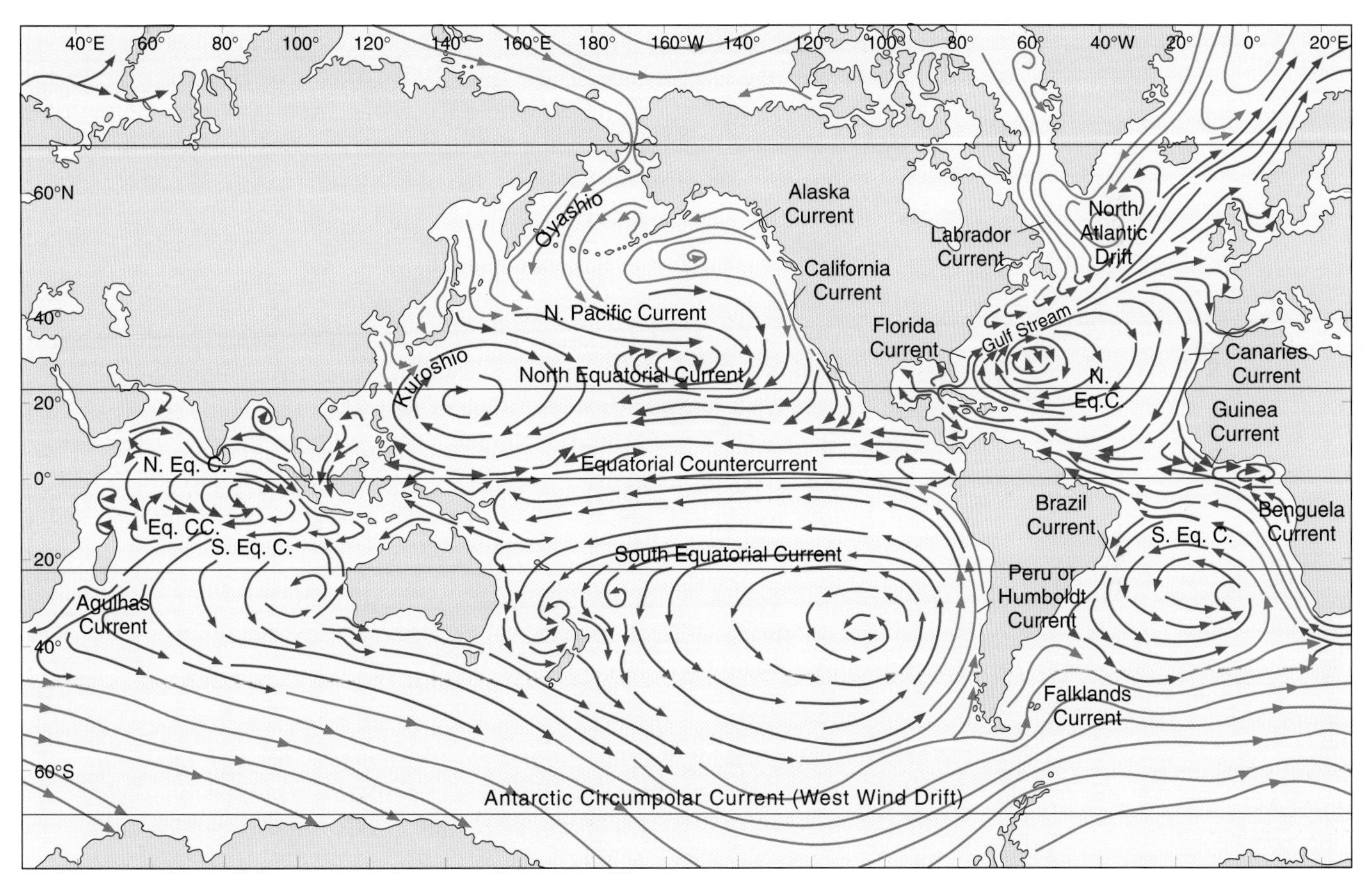

Figure 4.4 A schematic diagram of the surface currents of the Earth for northern summer. Red arrows represent warm ocean currents, and blue arrows represent cool currents. In some regions there are large seasonal differences.

❍ How different is Figure 4.4 to Major Rennell's map in Figure 4.3?

● Rennell's map is really rather good. He has managed to get the currents around the African coast virtually correct, and has also drawn a clockwise circulation in the Gulf of Guinea.

Figure 4.4 shows that the cold waters off the coast of Portugal mentioned in the previous section are now clearly part of a vast clockwise surface-circulation pattern in the North Atlantic, which consists of the Gulf Stream, the North Atlantic Current and the North Equatorial Current. This clockwise circulation is reflected in an anticlockwise pattern in the South Atlantic Ocean. In the Pacific Ocean a similar pattern can be seen, that is clockwise currents in the Northern Hemisphere and anticlockwise currents in the Southern Hemisphere. The main difference between the two oceans is that the circulation in the Pacific Ocean is orientated more along lines of latitude than those in the Atlantic Ocean. There is also an anticlockwise circulation seen in the Indian Ocean. In the higher latitudes of the Arctic and Antarctic the situation is somewhat different. In the Arctic there are a series of smaller closed-circulation patterns. This is contrasted in the Antarctic where there is a continuous current around the Antarctic continent called the Antarctic Circumpolar Current. In Section 4.2 we will explore what it is that makes the oceans move.

4.2 Wind forcing of the oceans

One of the main forces that moves the surface of the oceans is provided by the wind. The winds that we feel on land are caused by air rushing to regions of different atmospheric pressures. These winds also, of course, pass over the oceans. When the winds blow across the surface of the water, friction causes energy to be transferred from the wind to the water through a force called the **wind stress**. It has been found by experiment that the value of the wind stress is directly proportional to the square of the wind speed. Wind stress is usually given the Greek symbol τ:

$$\tau \propto W^2 \qquad (4.2)$$

where W is the wind speed. Equation 4.2 tells us that if the wind speed increases from $1\,\mathrm{m\,s^{-1}}$ to $4\,\mathrm{m\,s^{-1}}$ then the wind stress will increase by a factor of 16. The reason for this large jump in the wind stress is that when the winds are weak the surface of the ocean is relatively flat (Figure 4.5a) and there are few wave tops for the wind to actually push against. As energy is transferred from the wind to the ocean, surface waves develop, the surface becomes rougher and 'stretched' and so more of the surface is actually in contact with the wind (Figure 4.5b).

This increased surface area in turn leads to more energy being transferred to the ocean and the development of large surface waves. Because Equation 4.2 shows that τ is directly proportional to the square of the wind speed W, we can also write it as:

$$\tau = cW^2 \qquad (4.3)$$

Where c is the constant of proportionality. Unfortunately, because the wind increases the roughness of the sea-surface, the constant of proportionality c is not actually constant over the whole range of wind speeds found over the ocean.

The effect of the wind stress on the surface of the ocean is twofold: first it causes the surface of the water to move, and secondly some of the energy from the wind is transferred downwards into the water column through internal friction called **eddy viscosity**. Eddy viscosity is caused by friction between water molecules as they rub together. The result is that the eddy viscosity transfers momentum downwards and mixes up the surface waters so that the temperature, salinity and density have uniform properties (see Figures 3.8, 3.11 and 3.13). This is shown in Figure 4.6.

- ❍ What effect will increasing the wind speed have on the thickness of the layer of well-mixed water?
- ● The thickness of the well-mixed layer will *increase* because the surface waters will be mixed to a greater depth as more energy is supplied.

(a)

(b)

Figure 4.5 (a) A typical sea-surface when winds are low. The surface is flat and there are few wave tops for the wind to push against. (b) A typical sea-surface when winds are higher. There are large waves 'stretching' the surface of the ocean and many wave tops for the wind to push against.

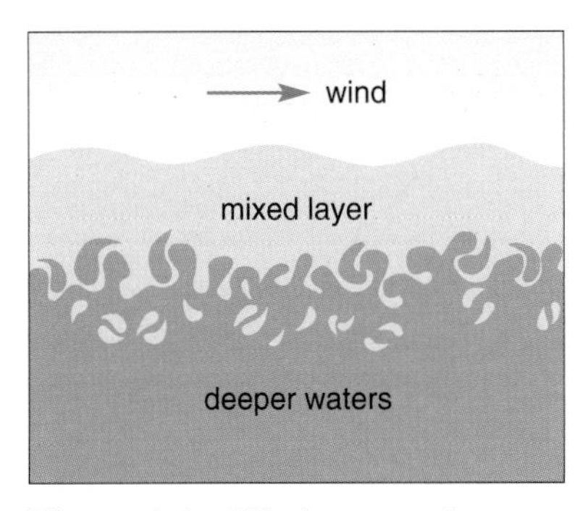

Figure 4.6 Winds across the surface of the ocean transfer energy downwards into the surface layers through eddy viscosity to create a surface layer with uniform physical properties. (Surface waters are light blue and the deeper waters dark blue.)

So when the wind blows across the oceans, the surface waters are mixed up and they start to move. But a key question is whether these surface waters actually drift in the direction of the winds. If the winds were the only force acting on the oceans, then they would. You have already seen in Block 2, Part 1 that when something is moving on the surface of the Earth it is affected by the Coriolis force. In the following text we shall look at whether this force also affects the waters of the oceans.

4.2.1 Ekman motion

The easiest way to see if the oceans move in the direction of the winds is to place a floating object on the water and see if it actually follows the wind direction. Something that does float rather well is ice, and in the Arctic and Antarctic the icebergs have been wonderful natural tracers of ocean currents. The early explorers to the Arctic soon noticed that the icebergs did *not* drift in the same direction as the winds. However, it wasn't until the early part of the 20th century that the Swedish mathematician Vagn Ekman discovered the reason for this departure, using data collected by fellow Scandinavian Fridtjof Nansen on his epic expedition to the Arctic.

Box 4.2 Fridtjof Nansen and the voyage of the Fram

Fridtjof Nansen (1861–1930) was a Norwegian Scientist who began his career as a zoologist before moving into ocean science. After leading the first successful expedition to cross Greenland in 1888 he decided that he wanted to explore the Arctic Basin to investigate the ocean circulation in this permanently ice-covered region. His reason was to see how it affected the Greenland and Norwegian Seas. Nansen became convinced that there must be an ocean current across the Arctic, as Siberian driftwood had been discovered on the coast of East Greenland. This conviction became stronger when wreckage from an American ship called the *Jeanette*, which had foundered in the Arctic ice north of Siberia, was also found in East Greenland. He raised the money to build a wooden ship, 123 feet in length, called *Fram* (Norwegian for 'Forward') that was designed to survive in the ice without being crushed. He left with 12 crew on board in July 1893. By September of that year the ship was frozen into the ice north of Siberia. In October, Nansen wrote in his diary:

> The ice is pressing and packing around us with a noise like thunder. It is piling itself into long walls, and heaps high enough to reach a good way up the *Fram*'s rigging; in fact it is trying its very utmost to grind the *Fram* into powder.

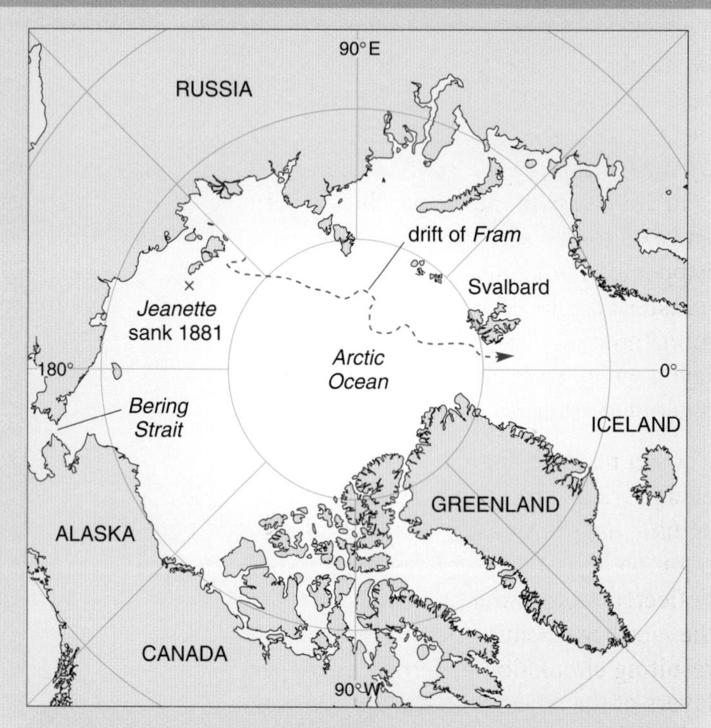

Figure 4.7 The drift of *Fram* across the Arctic Ocean, and the location of the sinking of the *Jeanette*.

Fram was well designed and as the ice crushed inwards the ship rose up above the ice. It was, however, the start of a voyage that did not finish until the spring of 1896 when the *Fram* was released near the island of Svalbard. Figure 4.7 shows the track of the *Fram* as she drifted across the Arctic Ocean.

As the ship slowly drifted whilst trapped in the ice, Nansen and the other crew members made a series of groundbreaking scientific measurements, which included measurements of the direction of the drift of *Fram* and the wind direction. These data were used by Ekman to derive the existence of the so-called Ekman Spiral (see main text). After his voyage, Nansen became a university academic and then a great statesman. He was awarded the Nobel Peace Prize in 1922. *Fram* also went on to great things — she was the ship Roald Amundsen used to reach Antarctica on his successful expedition to reach the South Pole in 1910. She can still be seen in a museum in Oslo.

When Ekman compared the direction of the drift of the *Fram* and sea ice to the wind direction, he soon noticed that, in general, the drift direction was 20–40° to the right of the wind direction. He explained this observation by developing a theory of wind-driven ocean currents.

He started with the theoretical idea of having an infinitely deep and infinitely wide ocean with no variations in density. Thus the ocean could be thought of as being made up of a series of infinite horizontal layers laid one on top of the other. As the wind blows across the surface of the ocean, the wind stress means that the surface will start to move. As the surface moves it is acted upon by the Coriolis force and so is deflected to the right (in the Northern Hemisphere). This moving surface layer is also acted upon by friction with its lower surface — the eddy viscosity — and so the layer underneath starts to move. Transfer of momentum by friction is an inefficient process and so the amount of energy transferred from layer to layer is greatly reduced. This means that when the forces are in balance, the speed of the second layer will be much less than the first layer. The balance of forces in the second layer means that it too is deflected to the right by the Coriolis force. This second layer is in contact with the deeper third layer and exactly the same processes are happening but again less energy is transferred. The eddy viscosity transfers momentum downwards to the third layer, which ends up moving, but again at a reduced speed.

The net result is that the amount of energy transferred downwards significantly decreases with each layer, but more importantly each layer will be deflected more and more to the right (in the Northern Hemisphere). This resulting circulation pattern in the upper layers of the ocean is called the **Ekman spiral** (Figure 4.8) as the influence of the wind decreases in each successive layer. The total depth of the frictional influence of the wind is called the **Ekman layer**.

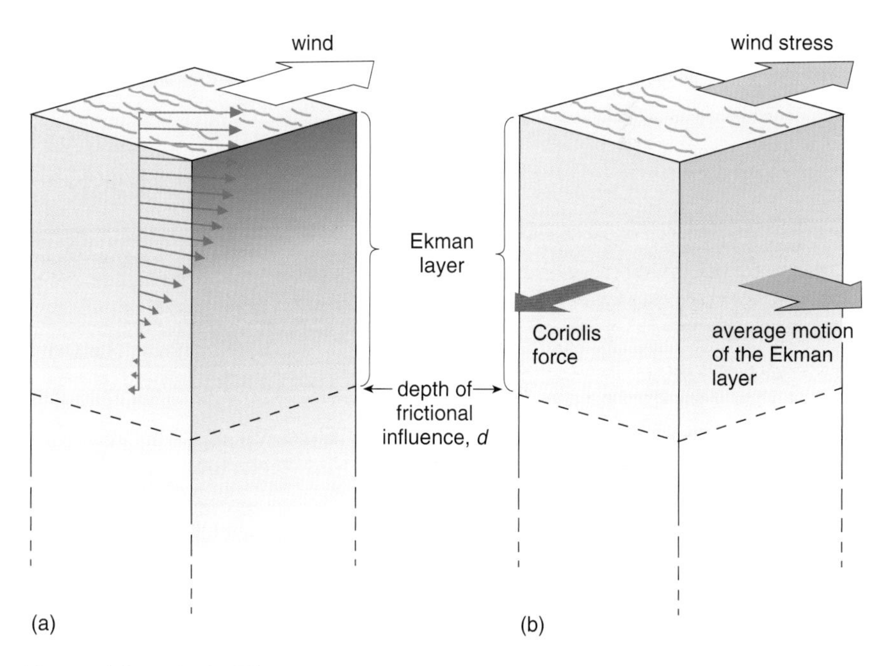

Figure 4.8 (a) The Ekman spiral pattern under ideal conditions in the Northern Hemisphere. Each successive layer of water is deflected to the right. The lengths of the arrows correspond to the strength and direction of the current. (b) The whole Ekman layer. The force from the wind is balanced by the Coriolis force. In the Southern Hemisphere the direction of the spiral and the average motion is reversed.

Ekman found that, in his ideal ocean, the direction of the surface current will be approximately 45° to the right of the wind direction in the Northern Hemisphere, and 45° to the left of the wind direction in the Southern Hemisphere. The magnitude of the Coriolis force on a particle of mass m and moving with a speed u is given by:

$$\text{Coriolis force} = m \times 2\,\Omega \sin\phi \times u \qquad (4.4a)$$

where the Greek letter omega, Ω, is the angular velocity of the Earth about its axis, and the Greek letter phi, ϕ, is the latitude and so ranges from 0° to 90° in the Northern Hemisphere and 0° to −90° in the Southern Hemisphere. In general the term $2\Omega \sin \phi$ is called the **Coriolis parameter** and is given the symbol f. This means we can re-write Equation 4.4a as:

$$\text{Coriolis force} = mfu \qquad (4.4b)$$

This is the force that acts on the surface layer of the ocean when it is under the influence of the wind on the rotating Earth. The speed of the surface current under the influence of the wind is given by:

$$u_0 = \frac{\tau}{\rho\sqrt{A_z f}} \qquad (4.5)$$

where u_0 is the speed of the surface current, τ is the wind stress (Equation 4.3), ρ is the density of the water, f is the Coriolis parameter, and A_z is a coefficient that represents the eddy viscosity.

- At a latitude of 75° N, what is the Coriolis parameter f given that $\Omega = 7.29 \times 10^{-5}\,\text{s}^{-1}$?

- $f = 2\Omega \sin \phi$

  Substituting $\Omega = 7.29 \times 10^{-5}\,\text{s}^{-1}$ and latitude $\phi = 75°$ N into this equation:

  $f = 2 \times 7.29 \times 10^{-5}\,\text{s}^{-1} \times \sin 75°$

  $f = 2 \times 7.29 \times 10^{-5}\,\text{s}^{-1} \times 0.9659$

  $f \approx 1.4 \times 10^{-4}\,\text{s}^{-1}$

Question 4.1

At a latitude of 40° N, what is the Coriolis parameter f given that $\Omega = 7.29 \times 10^{-5}\,\text{s}^{-1}$?

A westerly wind blows over the surface of the ocean with a speed of $10\,\text{m}\,\text{s}^{-1}$ at 40° N. Assuming the wind stress τ is $0.1\,\text{Nm}^{-2}$, the eddy viscosity A_z is $0.1\,\text{m}^2\,\text{s}^{-1}$ and the density of water is $1000\,\text{kg}\,\text{m}^{-3}$, what is the speed of the wind-generated current according to Equation 4.5?

In which direction does the wind-generated current u_0 flow?

The limitations of Ekman's theory are that the oceans are not infinitely deep, and the coefficient of eddy viscosity A_z is not constant with depth. A_z actually changes by several orders of magnitude. It is generally much higher near the surface and decreases as depth increases.

The most important thing about Ekman's theory is not that the wind-generated surface currents are deflected by the Coriolis force but that there is a surface layer of water that is not necessarily the same thickness as the mixed layer. This layer is influenced by the wind and moves with a mean current over the depth of the Ekman layer, to the right of the wind in the Northern Hemisphere and to the left in the Southern Hemisphere. Ekman went on to show that the average of the current over the depth of its influence, given the symbol $\bar{u}$ (the bar over the u means average), is given by:

$$\bar{u} = \frac{\tau}{D\rho f} \qquad (4.6)$$

where D is the depth at which the wind driven current is directly opposite the surface current. It is sometimes called the depth of frictional influence (the other symbols have already been defined). This mean current $\bar{u}$ is called the **Ekman Drift**.

- ❍ Will the depth of frictional influence increase if the value of the eddy viscosity A_z increases?
- ● Yes. The eddy viscosity A_z is the coefficient telling us how much energy is transferred downwards. If it increases, the depth of frictional influence D will increase.

Under ideal conditions and using a typical value for A_z, the value of D is approximately 40 m for high-latitude polar regions. It increases in mid-latitudes to approximately 50 m and, close to the Equator, theoretically the value of D rapidly approaches infinity. In the real oceans the value of D is strongly influenced by the thickness of the mixed layer. This means it is largely dependent on the *time of year* and *wind speed* and typically varies from tens of metres up to 200 m. When the mean current $\bar{u}$ is multiplied by the thickness D, we can calculate the total transport of water to the right (in the Northern Hemisphere) due to the Ekman drift, and this is called **Ekman transport**. We shall see that Ekman transport is very important for understanding the general circulation of the oceans.

Armed with this basic theory of the wind-driven surface currents we can now look at a map of the global surface winds and think about their effect on the oceans. We have already seen in Block 2, Part 1 that there is a mean global wind pattern caused by the interaction of incoming solar radiation and the land-masses of the Earth. This mean global surface-wind pattern is shown in Figure 4.9.

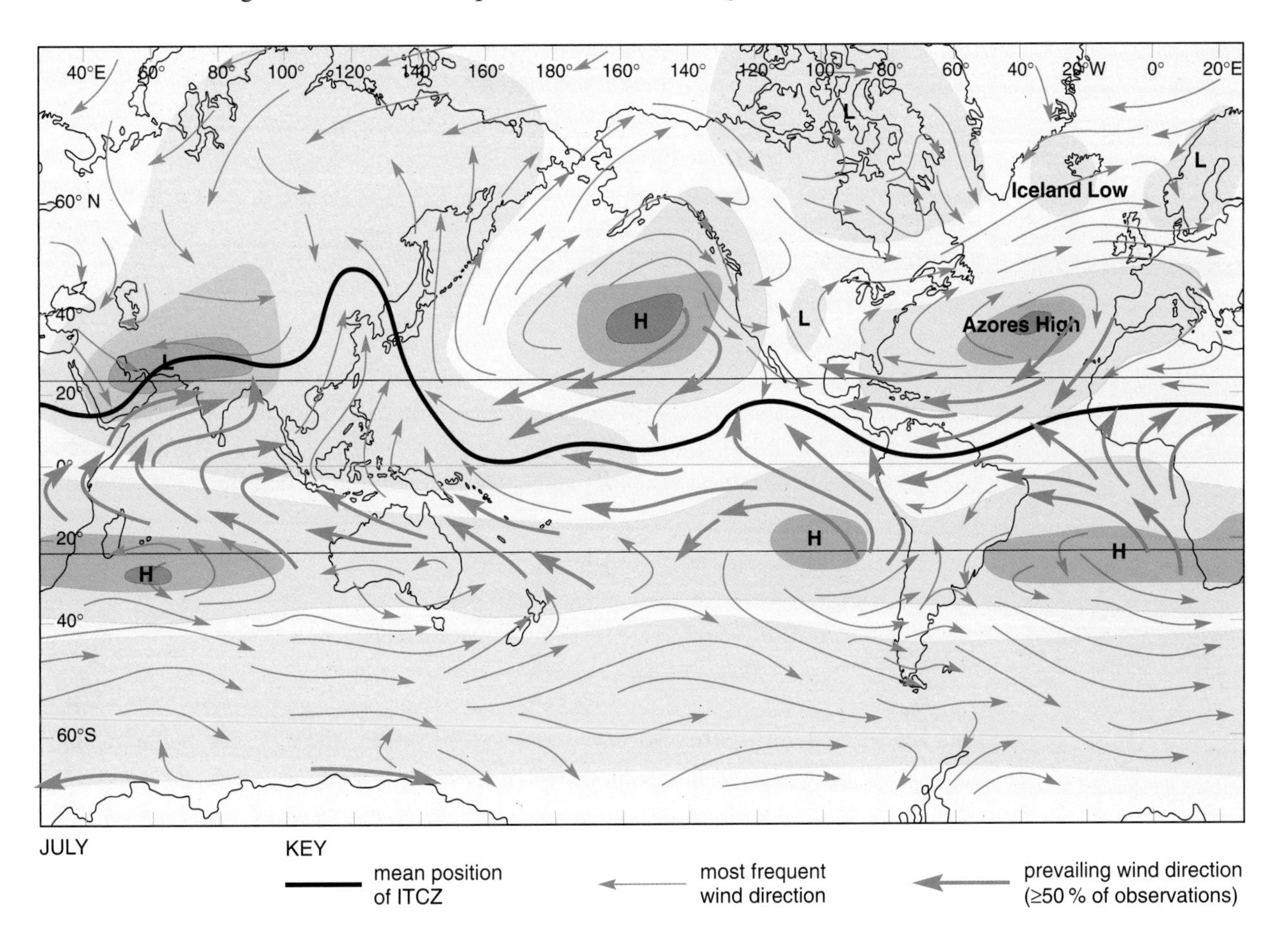

Figure 4.9 The mean global surface-wind pattern and the average position of the Intertropical Convergence Zone (ITCZ) in July.

Figure 4.9 can be compared with the map of mean surface currents in Figure 4.4. To a certain extent the maps compare rather well. The wind-driven component of the ocean circulation appears to follow the mean winds in general. Where surface currents join in continuous circulation patterns like in the North Atlantic the effect of the Ekman drift is striking, as described below.

4.2.2 Divergence and convergence

If we consider a region with closed surface-wind circulation such as the North Atlantic Ocean (Figure 4.9) there is a clockwise wind pattern around an atmospheric high pressure. This clockwise circulation is repeated in the sea-surface current map in Figure 4.4. Such a closed surface-water circulation is called a **gyre**, and this particular closed pattern in the North Atlantic Ocean is called the North Atlantic gyre.

- What will be the effect of Ekman drift in the centre of the North Atlantic gyre?
- The Ekman drift will cause a slow movement of water into the centre of the gyre.

What actually happens is that the drift of water into the centre of the gyre causes a surface convergence that pools warm water in the centre of the gyre and raises the surface of the ocean by approximately one metre.

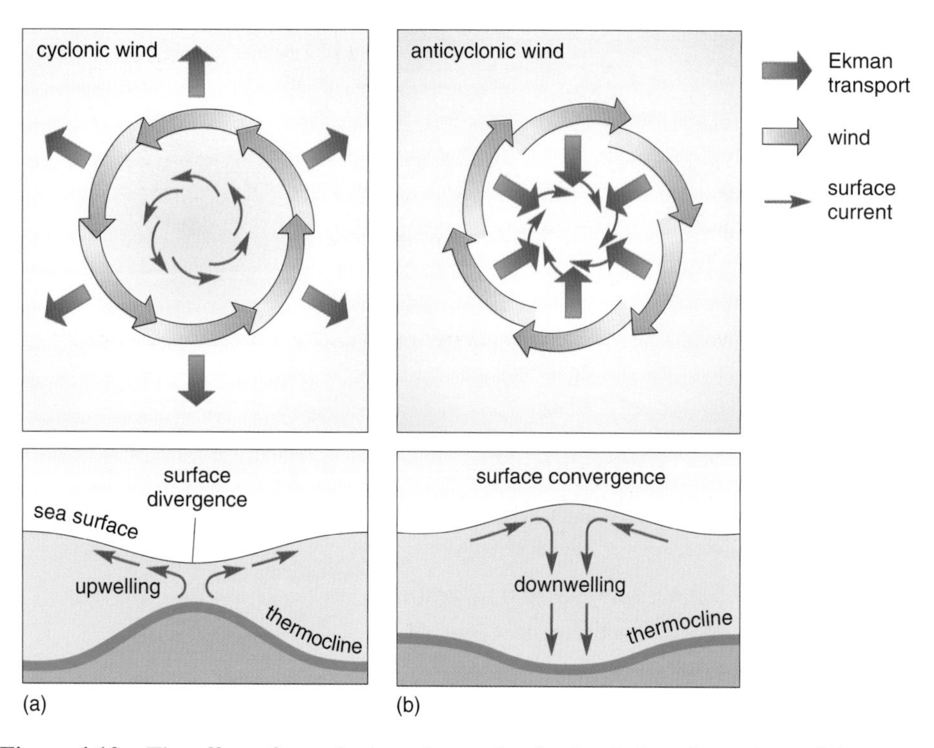

Figure 4.10 The effect of a cyclonic and an anticyclonic wind on the surface of the ocean. (a) In a cyclonic wind there is a divergence at the surface. This divergence depresses the surface of the ocean and raises water from beneath the thermocline towards the surface (upwelling). (b) In an anticyclonic circulation the surface waters converge and this pushes the sea-surface upwards and depresses the thermocline (downwelling).

Question 4.2

What will happen to the level of water in the centre of a gyre in the Northern Hemisphere where the circulation of the winds and surface currents is anticlockwise?

A clockwise circulation is called anticyclonic circulation, and an anticlockwise circulation is called cyclonic circulation. Where there is an anticyclonic circulation there is a convergence of surface waters. In a cyclonic circulation there is a divergence of surface waters. When there is a convergence or divergence at the surface of the ocean, there is a corresponding effect beneath the surface. This is shown in Figure 4.10.

4.3 Thermohaline forcing of the oceans — the latitudinal circulation

The surface-circulation patterns (Figure 4.4) and the resulting Ekman drift can be compared with the SST (Figure 3.7) and the sea-surface salinity (Figure 3.10). There are large volumes of water in the mixed layers where both the temperature and salinity are relatively constant. Such a large volume of water with relatively constant temperature and salinity is called a **water-mass**. We showed in Section 3 that the surface temperature and salinity in a particular geographical region are determined by the regional climate. This allows us to name a water-mass after the region from where it got its temperature and salinity characteristics. Figure 4.11 shows the names of the more commonly defined surface water-masses.

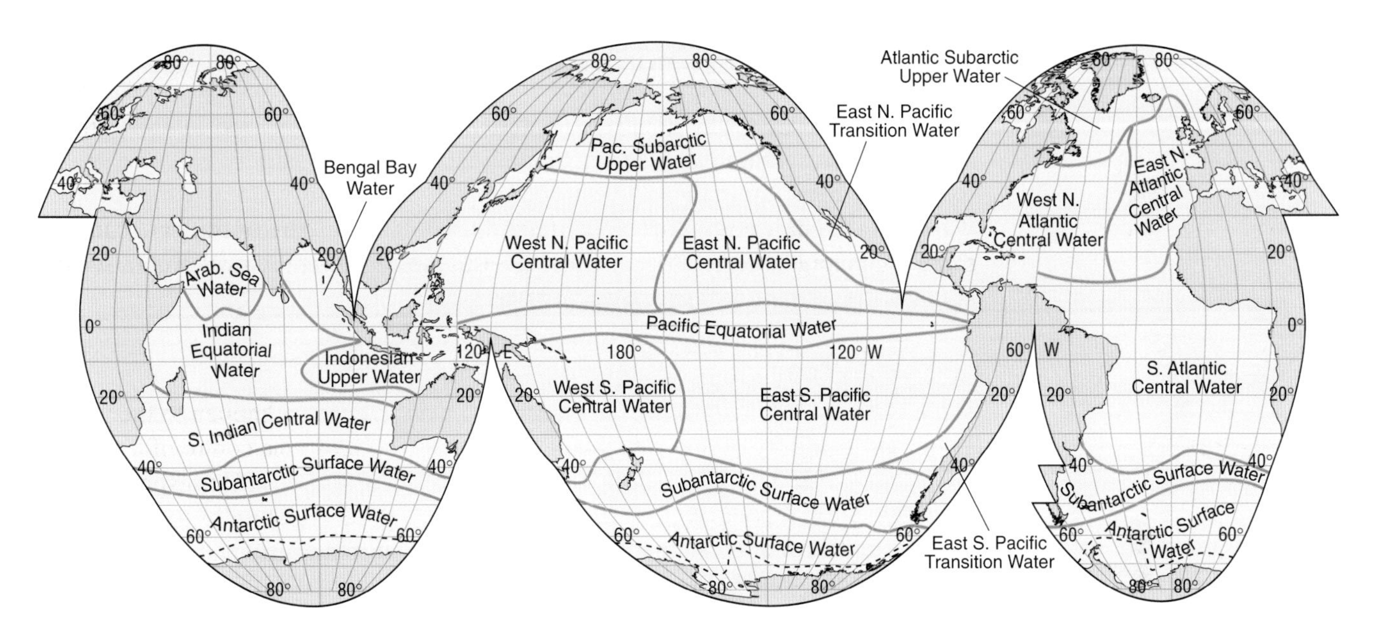

Figure 4.11 The global distribution of the upper-ocean water-masses. The boundaries between the different water-masses are not sharp like the lines on this graph, but are more diffuse.

Question 4.3

Do you think that all the surface water-masses shown in Figure 4.11 will have the same density?

In Figure 3.14 we saw that the density of seawater is a function of both temperature and salinity. We also saw in the answer to Question 3.4 that for the usual range of oceanic salinities there is only a small region of this plot that is typical of oceanic salinities. Figure 3.14 can be redrawn with an x-axis showing salinity, and a y-axis showing temperature for the range of temperatures and salinities that we generally encounter in the oceans, and with density units as contours. Such a plot is shown in Figure 4.12 (overleaf).

By using Figure 4.12 we can see that water-masses that have different temperatures and salinities will have different densities. In the polar regions (Figures 3.7 and 3.10), sea-surface temperatures are approximately −1 °C and salinity is about 34, so

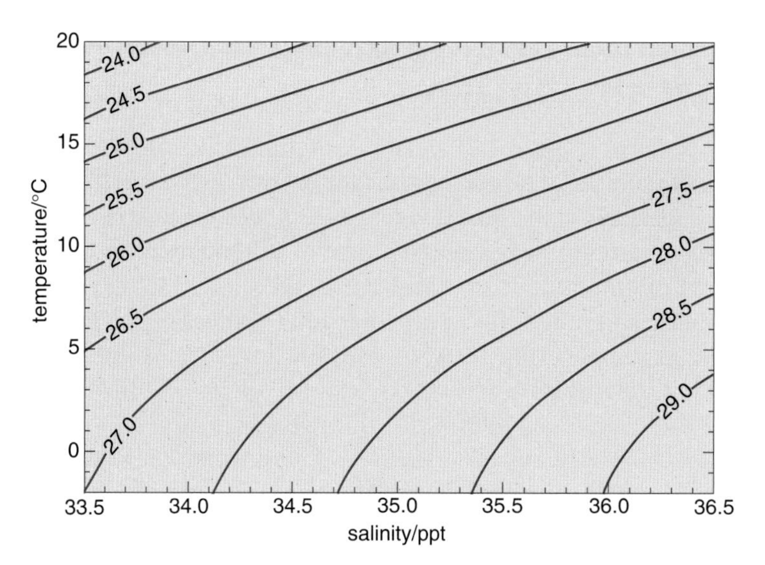

Figure 4.12 A temperature and salinity plot showing the range of typical temperatures, salinities and density anomalies found in the oceans.

the density anomaly will be approximately 27.4 kg m^{-3}. In the mid-latitude regions sea-surface temperatures are approximately 20 °C and the salinity is 35, so the resulting density anomaly will be approximately 24.9 kg m^{-3}. Clearly the surface waters in the polar regions are much more dense than the surface waters in the more temperate latitudes. When water-masses with different densities come into contact, the densest will sink to the bottom.

Question 4.4

What is the density anomaly of a water-mass with a temperature of approximately 15 °C and a salinity of approximately 35.0?

In certain special polar regions, for example the Greenland Sea, there is a cyclonic circulation that will mean that the thermocline is pulled towards the surface (Figure 4.10). In these regions the surface water is always cold, but when ice is formed, salt is rejected from the ice matrix and this increases the salinity of the thin surface layer. When the surface waters are both cold and salty, as more salt is added to the thin layer its density will be greatly increased (Figure 4.12). Eventually, if enough salt is added, the density of the surface waters will increase to the point where they will sink through the thermocline. The water in these regions is so dense that it sinks to great depths and is called **deep water**. Deep water formed in the North Atlantic is called North Atlantic Deep Water (NADW).

Figure 4.13a shows the areas where deep water is formed in the North Atlantic Ocean.

The three areas in the Arctic where deep water is generally formed are the Greenland and Norwegian Seas, and the Labrador Sea. All three areas have the characteristics we have talked about — they have cyclonic circulations coupled to regions where sea-surface temperature is very cold and ice is sometimes formed. After formation in the Greenland and Norwegian Seas the deep water flows through deep channels in the ridge system between Greenland and Scotland, such as the Denmark Strait, and enters the North Atlantic Ocean. These dense waters are joined by a deep water formed under similar conditions in the Labrador Sea. In the Antarctic similar conditions exist in gyres close to the Antarctic continent. The most important regions for these processes are the Weddell Sea at the bottom of the

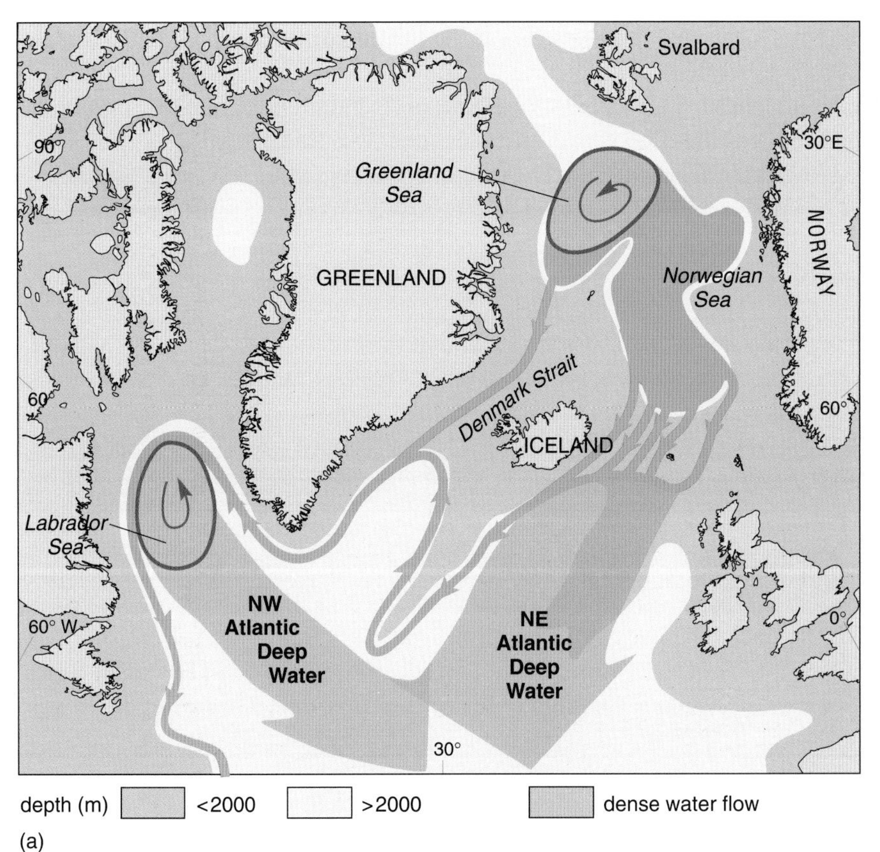

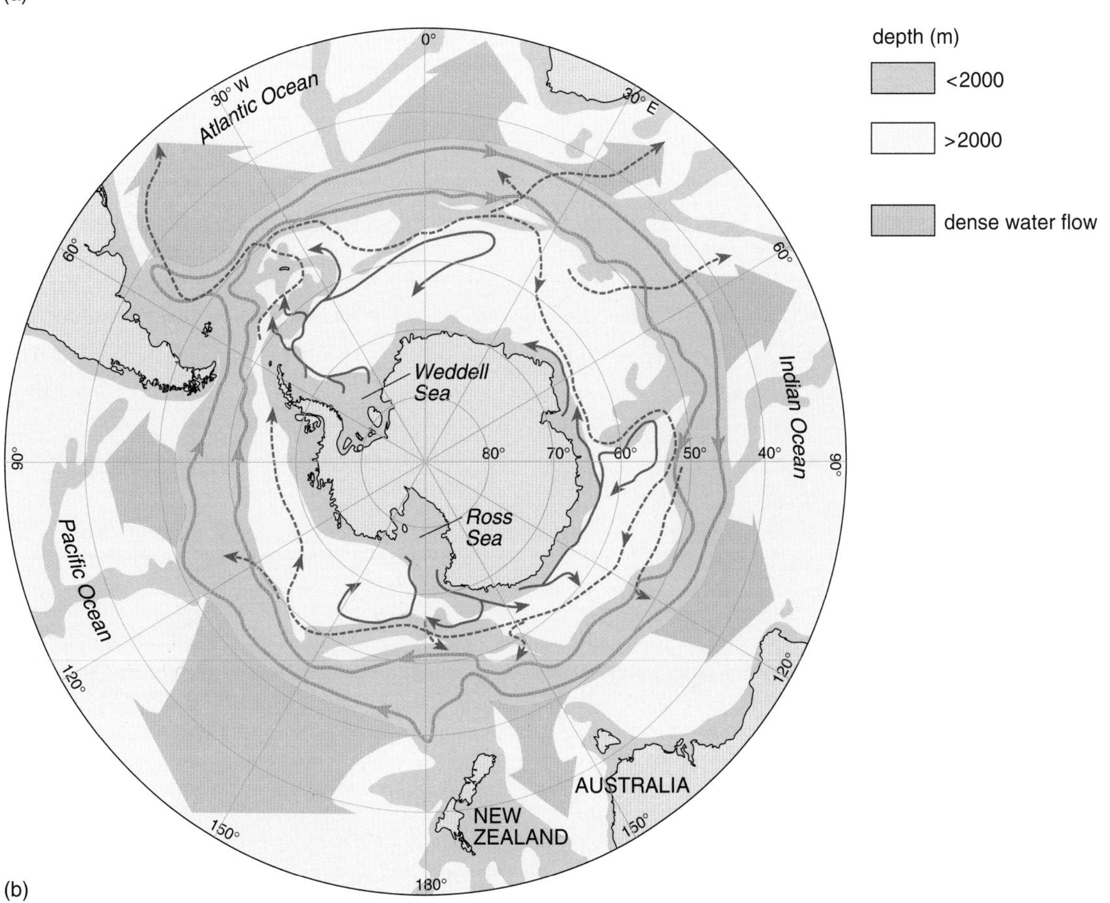

Figure 4.13 Deep and bottom water formation in the Arctic and Antarctic. The solid and dashed blue lines represent the direction of ocean currents. Shades of turquoise represent the flow of deep ocean currents (the darker the shade of turquoise the deeper (and denser) the current). (a) In the Arctic the deep water is principally formed in the Greenland and Norwegian Seas, and also in the Labrador Sea. This dense water sinks down and spreads out through the Atlantic Ocean. (b) In the Antarctic, bottom water is formed in cyclonic gyres close to the Antarctic continent. This dense water then flows away from the Antarctic continent to fill up the deep ocean basins of the Earth.

South Atlantic Ocean and the Ross Sea at the bottom of the South Pacific Ocean. In these regions there is a cyclonic circulation (clockwise) because it is in the Southern Hemisphere. This circulation again pulls the thermocline towards the surface and reduces the thickness of the surface layer. Because of the different salinities present at the surface of the Antarctic waters, when ice forms on the surface of the ocean and salt is added to the surface layers, the surface waters can become dense enough to sink all the way to the sea floor. Once at the sea floor the dense water floods out into the other oceans of the world. As the water can sink all the way to the sea floor it is called **bottom water**. As in the North Atlantic the water is named after the region in which it is formed, in this case of course the Antarctic. This means the bottom water is called Antarctic Bottom Water (AABW).

- What will happen when the dense NADW from the Arctic finally meets the denser AABW from the Antarctic?
- The AABW water will stay close to the sea floor as it is more dense and the NADW will move up over it to fill the mid-depths of the ocean.

A two-dimensional schematic diagram of the vertical circulation pattern of the Atlantic Ocean is shown in Figure 4.14. The dense AABW (coloured blue) forms in the clockwise circulating cyclonic gyres of the Antarctic. This sinks down to the bottom of the South Atlantic Ocean and spreads northwards along the sea floor. The colourless water-mass represents water that is formed by the climatic conditions in the mid-latitudes of the Southern Hemisphere that is not as dense as AABW. In the Arctic the NADW is formed in the anticlockwise circulating gyres. It crosses the ridges between Greenland and Scotland to flow southwards. When it reaches the water-masses from the southern ocean, the NADW is more dense than the colourless water-mass, and less dense than the AABW. The result is that the NADW is sandwiched between these two water-masses from the Southern Ocean.

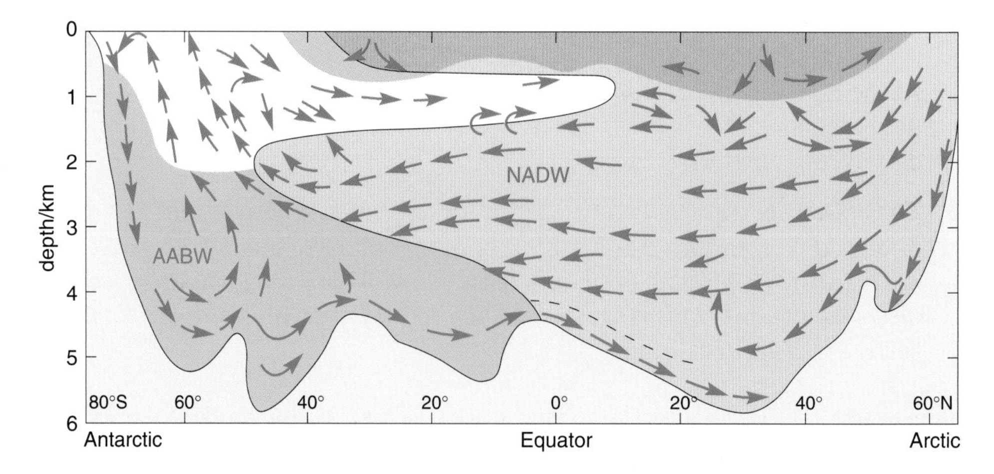

Figure 4.14 A two-dimensional cross-section of the circulation pattern in the Atlantic Ocean. AABW that is formed in the Antarctic is dense enough to fill up the deep basins of the South Atlantic. The NADW is less dense and so it rises up over the AABW and fills up the mid-depths of the South Atlantic where the water-masses meet. Blue arrows represent the movement of the different water-masses; different colours represent different water-masses.

The schematic diagram in Figure 4.14 can be compared with the actual temperature and salinity distributions that have been measured in the Atlantic Ocean (Figures 3.9 and 3.12). The range of temperatures in any vertical profile can now be seen to have arisen from cold waters sinking at high latitudes and spreading throughout the oceans. This explains the 'U' shape of the isotherms, with cold water at the surface in high latitudes and deep water in the equatorial and low latitude regions. In the vertical section of salinity across the Atlantic Ocean (Figure 3.12), the isohalines that revealed an apparent clockwise circulation can be seen to result from the process shown in Figure 4.14. The vertical distribution of salinity in the oceans is therefore a result of the density differences between different water-masses that have been formed under different climatic conditions.

4.4 The global ocean circulation

We have seen that the climate in different parts of the world can create conditions where different density waters find their own 'level' and so drive a vast circulation across the Atlantic Ocean. This circulation is apparent from the temperature and salinity profiles. Temperature and salinity are called 'conservative' properties of seawater. Once a parcel of seawater gets its particular temperature and salinity at the sea-surface, when it sinks, the temperature and salinity will not change unless the parcel of water mixes with another parcel that has a different temperature and salinity. There are many other things we can measure that are dissolved in the seawater that are not conservative but still reveal a lot about the circulation of the ocean. One of the more common measurements is dissolved oxygen. When seawater is at the surface and in contact with the atmosphere, oxygen dissolves in it. The amount of oxygen that can be dissolved in water is strongly dependent on temperature, with the colder waters being able to dissolve much more oxygen than warmer waters. Oxygen is a non-conservative tracer because once it is away from the surface of the ocean it can be used up by biological processes, so the amount can decrease without mixing with any other seawater.

❍ At the Equator in Figure 4.14 would you expect the amount of oxygen dissolved in the NADW at 2 km depth to be higher or lower than the amount dissolved in the surface water?

● The equatorial surface water is very warm so it will have a low amount of dissolved oxygen. The NADW at 2 km depth will have been at the surface in the higher latitudes and, because it is colder, will have a higher amount of dissolved oxygen. This means the amount of oxygen will increase from the surface to a depth of 2 km.

By mapping out both conservative and non-conservative properties of seawater, scientists have deduced that the circulation in the oceans is not just from high to low latitudes as demonstrated above, but also between different ocean basins. The different temperatures and salinities caused by regional climatic variations create surface water-masses with a range of different densities, as shown in Figure 4.11. The resulting density-driven currents have been mapped out across the Earth in a highly schematic way in a pattern called the 'oceanic conveyor belt', as shown in Figure 4.15.

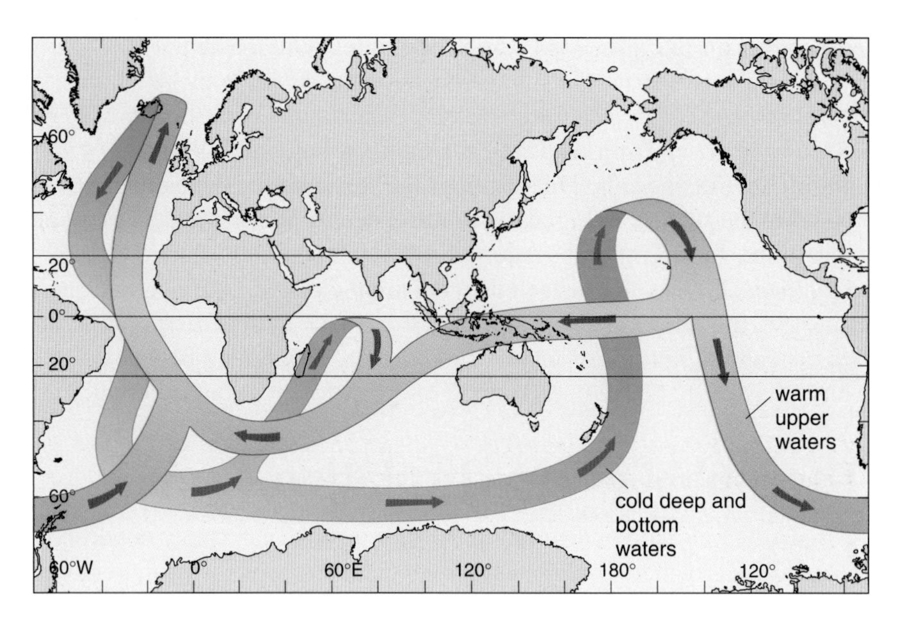

Figure 4.15 The 'oceanic conveyor belt': the red/orange arrow represents circulation in the upper 1 km of the ocean; the blue arrow represents water below this depth.

The concept in Figure 4.15 first arose in the 1980s and is constantly being updated as our knowledge of the oceans improves. We have seen in the Atlantic Ocean that the actual detail is much more complicated, and Figure 4.15 is really no more than a tool to help us think about the way the global ocean circulates. Putting that aside, it does help us to understand the way the circulation of the ocean moves heat around the planet. In general terms surface water is heated up in the Pacific Ocean and it flows both westwards carrying heat into the Indian Ocean, and southwards towards Drake Passage and into the South Atlantic. The warm branch of water in the Indian Ocean flows around the coast of Africa and also enters the South Atlantic. The two warm branches can be thought of as combining and flowing northwards into the North Atlantic where they supply the heat that gives Western Europe such a temperate climate. In the high-latitude regions of the North Atlantic the waters are cooled. We have seen that when this is combined with the result of the $E-P$ balance the waters become dense enough to sink and leave the surface to flow back southwards. Eventually the cooler branch returns to the Pacific Ocean and the cycle begins again. It probably takes several hundred years for an individual parcel of water to go completely around this 'conveyor belt' circulation system. The schematic diagram in Figure 4.15 shows that there does not seem to be any of the dark-blue deep water formed in the Pacific Ocean. This is because of the combination of the shape of the ocean basins shutting off access to the very high latitudes in this ocean, and the climatic conditions.

Question 4.5

What could be the effect of reduced deep-water formation in the North Atlantic Ocean on the climate of Western Europe?

4.5 Summary of Section 4

1 The most obvious evidence that the oceans 'move' is through driftwood being washed up in places where it could never have grown. The ocean currents were mapped by early scientists and used to reduce passage times for ships.

2 The wind exerts a force on the ocean through the frictional force of wind stress. This force is proportional to the square of the wind speed.

3 The effect of the wind stress on the surface of the ocean is passed down through the water column through eddy viscosity, which is a result of friction between the water molecules. This transfers energy into the water column.

4 Because the sea is frictionally coupled to the rotating Earth, the Coriolis force is important. In the Northern Hemisphere this deflects the wind-driven water to the right. As momentum is transferred downwards, successive layers move at a slower velocity and are deflected more and to the right. This creates the Ekman spiral.

5 The net result of the Ekman spiral in the Northern Hemisphere is to cause a slow transport of water to the right of the wind; in the Southern Hemisphere it causes a slow transport of water to the left of the wind.

6 Large volumes of water formed in geographical areas of the ocean with uniform temperature and salinity are called water-masses. The water-masses are named after the geographic region in which they are formed and can have a wide range of densities due to their different temperature and salinity values.

7 Some water-masses are dense enough when they are formed to sink away from the surface to form deep and bottom water. These water-masses of different densities drive a vertical circulation pattern throughout the Atlantic Ocean.

8 The processes that drive a circulation pattern throughout the Atlantic Ocean also operate in the other oceans of the world. The differing densities of water have set up a global circulation pattern that has been drawn schematically as an 'oceanic conveyor belt'. This conveyor belt distributes large amounts of heat around the Earth.

Learning outcomes for Section 4

After working through this section you should be able to:

4.1 Explain how we know that the oceans circulate.

4.2 Explain why wind stress is proportional to the square of the wind speed.

4.3 Describe how the Ekman spiral evolves.

4.4 Explain how the Ekman transport will affect a cyclonic and an anticyclonic gyre in the Northern Hemisphere.

4.5 Recognize that water-masses formed under differing climatic conditions will have different temperatures and salinities, and so different densities.

4.6 Explain how the vertical circulation pattern arises in the Atlantic Ocean.

4.7 Describe the main features of the oceanic conveyor belt.

4.8 Suggest possible consequences of changes in the oceanic conveyor belt.

5 The oceans, carbon, and climate change

We have seen in the previous four sections that the oceans are a complicated physical system that can have a significant effect on the climate. Of course one of the main points we have not yet mentioned is, as the oceans cover >70% of the Earth's surface, they naturally provide a vast habitat for life on Earth.

There is also growing evidence that the increase of carbon dioxide in the atmosphere (Figure 5.1) is changing the climate of the Earth. In this final section we shall see the potential importance of the oceans on the future of our planet.

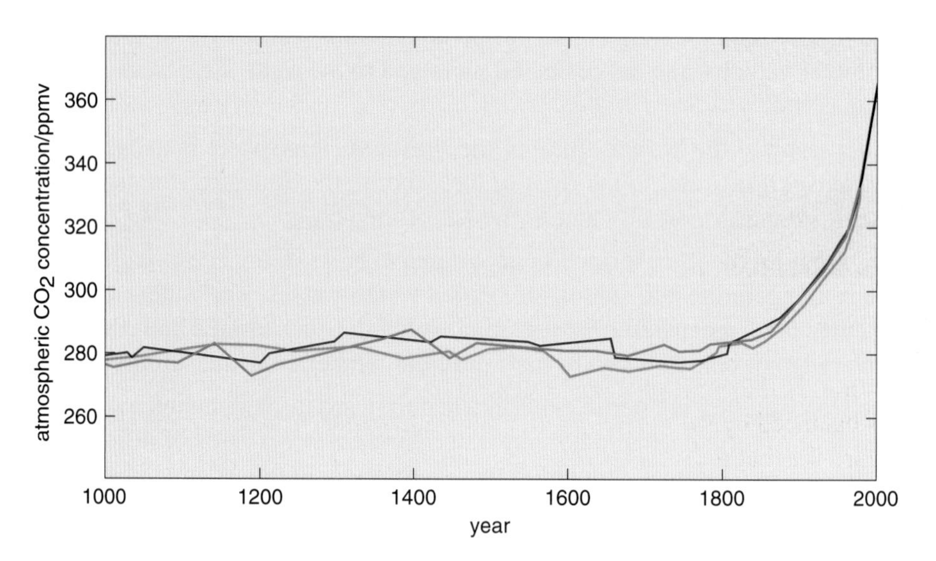

Figure 5.1 The atmospheric concentration of carbon dioxide over the last millennium. Note how rapidly this has increased since the year 1800. The coloured lines represent measurements using different techniques that have been compiled to make a continuous record. (Intergovernmental Panel on Climate Change, *Climate Change 2001.*)

5.1 The oceans and the carbon cycle

In Section 4 we noted that oxygen plays a role in biological processes. Another important gas in biological processes is, of course, carbon dioxide (CO_2). Just like oxygen, carbon dioxide also dissolves in the oceans.

Carbon dioxide plays the opposite role to oxygen in that it is utilized in photosynthesis and released by respiration. Phytoplankton, which are very small plants, carry out photosynthesis in the oceans. They generally live in the mixed layer (Figures 3.8, 3.11 and 3.13) close to the surface where there is lots of sunlight to provide the energy for photosynthesis. Oxygen is released during this process.

$$6CO_2 + 6H_2O \xrightarrow{\text{sunlight}} C_6H_{12}O_6 + 6O_2 \qquad (5.1)$$

Most of the CO_2 that is converted into organic carbon is released back into the ocean as the plants die and dissolve. Some organic carbon sinks to the sea floor and is trapped in the ocean sediment for long periods of time. Though there are several sources and sinks of carbon in the ocean, they are never in balance. We have already seen in Figure 5.1 that atmospheric CO_2 has increased rapidly in the last century. A schematic diagram of the natural carbon cycle is shown in Figure 5.2.

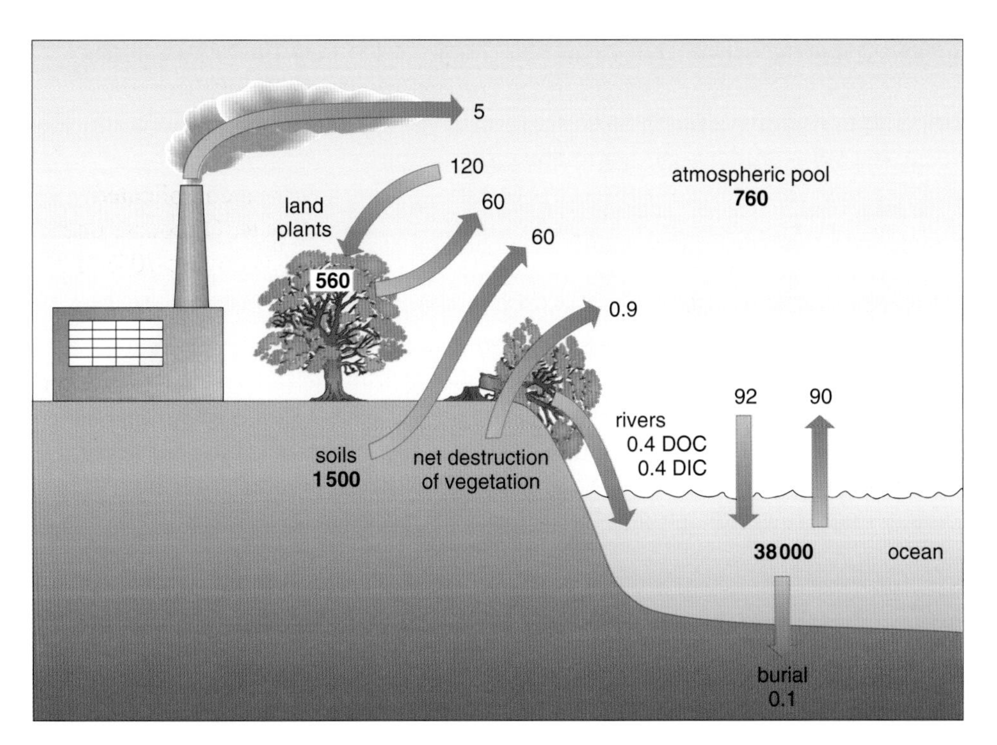

Figure 5.2 The present-day global carbon cycle. The sizes of all reservoirs (bold values) are expressed in units of 10^{12} kg C and annual fluxes in units of 10^{12} kg C yr^{-1}. DOC is dissolved organic carbon, DIC is dissolved inorganic carbon.

You can see in Figure 5.2 that the carbon stored in the oceans ($38\,000 \times 10^{12}$ kg C) dwarfs all of the other components of the system, and is 50 times larger than the carbon stored in the atmosphere. The carbon released by soils, and from land plants, is balanced by the carbon taken up by land plants. You may have also noted from Figure 5.1 that the amount of carbon dioxide in the atmosphere is increasing. The Intergovernmental Panel on Climate Change (IPCC) has estimated that in the 1980s about 35% of this increase was actually absorbed by the oceans.

Question 5.1

If the carbon released to the atmosphere in the 1980s was 5.5×10^{12} kg per year, how much of this was absorbed by the ocean per year? Will it be significant?

Clearly the oceans are a very large reservoir for carbon. You can see in Equation 5.1 that the process of photosynthesis removes carbon dioxide from the atmosphere, but what would happen if the amount of photosynthesis was greatly increased? Some people have suggested that we could reduce the increase in the amount of CO_2 in the atmosphere, and hence the greenhouse effect, by increasing the amount of carbon transferred to the ocean. But could it be so simple?

5.2 Future climates

The IPCC have predicted that the global increase in atmospheric CO_2 (Figure 5.1) will result in a predicted average global surface-temperature rise of between 1.4 °C and 5.8 °C between 1990 and 2100. This is on top of the already recorded change of approximately 0.6 °C in the 20th century. Such temperature rises will, for example, modify the sea-ice cover in the polar regions and the resulting average albedo. As the albedo changes it is likely that more solar energy will be absorbed in the polar regions, and water-mass properties will change. The IPCC have suggested that the oceanic conveyor belt (Figure 4.15) may slow down in the next 100 years. Of course changes in the conveyor belt and atmospheric temperature may drive much more significant transports in carbon. The truth is we are not yet confident enough to make predictions.

If we look back to our original example in Section 1, which compared the climate of two locations (Figure 1.2), we noted that Bergen had winter temperatures of up to 6 °C warmer than Halifax. As you already know, this heat is supplied as part of a global oceanic circulation. To predict the future climate of the Earth we need to calculate the implications of the predicted temperature rise of the global surface temperature when combined with the reduction in heat transport if the oceanic conveyor belt slows down. It has often been assumed that global warming will make northern Europe much warmer, but if the conveyor in Figure 4.15 was 'switched off', we may end up with temperatures that are actually much lower than today.

5.3 Summary of Section 5

1 In the last century the amount of carbon dioxide in the atmosphere has rapidly increased.

2 The amount of carbon stored in each component of the natural carbon cycle shows that oceanic carbon dwarfs all other reservoirs.

3 The oceans can remove carbon from the atmosphere by photosynthesis.

4 The average global surface temperature is predicted to rise in the next 100 years, and the oceanic conveyor belt is predicted to slow down. The result of these two effects on regional climate is not well understood.

Learning outcomes for Section 5

After working through this section you should be able to:

5.1 Describe the relative importance of the ocean and atmospheric carbon reservoirs.

5.2 Describe how carbon can be taken up by the ocean through photosynthesis.

5.3 Describe why it is so difficult to predict the future climate of the Earth.

Answers to questions

Question 2.1

The ship hears the echo after 6 s. This is equal to $2t$. So the time for the pulse to reach the sea floor is calculated as follows:

$$t = \frac{6\ \text{s}}{2}$$

$$t = 3\ \text{s}$$

The depth of the ocean = time for pulse to reach sea floor × speed of sound in seawater

$$= 3\ \text{s} \times 1500\ \text{m s}^{-1}$$

$$= 4500\ \text{m}$$

Therefore the depth of the sea floor at this point is 4500 m.

Question 3.1

The water at the surface of the pool will cool and become less dense. This means that it will not sink away from the surface, resulting in a cool surface layer with warmer, more dense water beneath.

Question 3.2

The main reason that the SST does not match up with the average incoming annual solar radiation is that the ocean is not static. Off the coast of Portugal, the cooler waters from the northwest are circulating in a clockwise direction bringing cooler waters south. Other reasons that the two do not match include the fact that we are comparing an annual field in Figure 3.6 with a image from just 4 days' data in Figure 3.7 and so seasonal effects are not shown.

Question 3.3

If you look at the poster map showing bathymetry, you can see that in the Arctic there is only one deep entry and exit to the Arctic Ocean between Greenland and Scotland. This exit is blocked to a depth of about 1000 m by a series of ocean ridges between Greenland and Scotland, so the cold water formed in the Arctic cannot cross into the Atlantic in the same way that the cold waters formed at the surface in the Antarctic cannot cross the Mid-Atlantic Ridge (see Figure 3.9).

Question 3.4

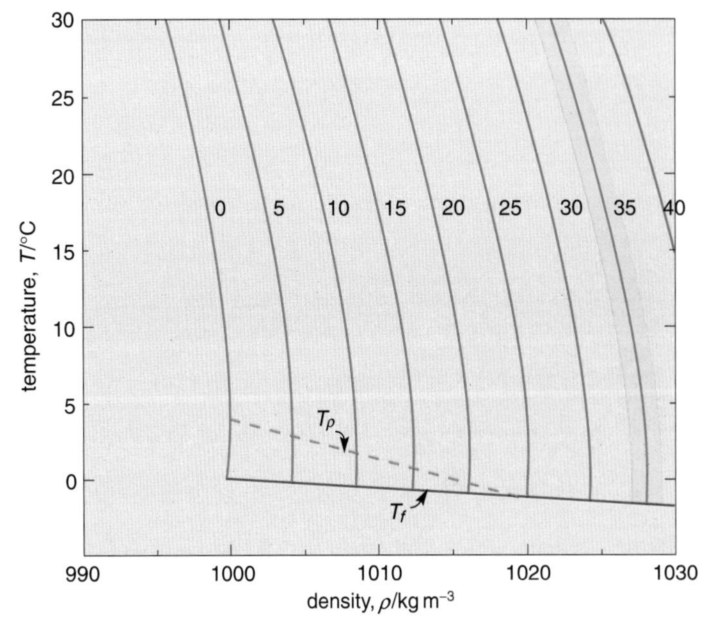

Figure 3.15 Relationship between density and temperature for different values of salinity. Shaded areas represent the region of densities expected in the oceans.

Question 4.1

$f = 2\Omega \sin\phi$

Substituting $\Omega = 7.29 \times 10^{-5}\ s^{-1}$ and latitude $\phi = 40°$ N into this equation:

$f = 2 \times 7.29 \times 10^{-5}\ s^{-1} \times \sin 40°$

$f = 2 \times 7.29 \times 10^{-5}\ s^{-1} \times 0.6428$

$f \approx 9.4 \times 10^{-5}\ s^{-1}$

Substituting appropriate values in Equation 4.5:

$$u_0 = \frac{0.1}{1000\sqrt{0.1 \times 9.4 \times 10^{-5}}}$$

$$u_0 = \frac{0.1}{1000\sqrt{9.4 \times 10^{-6}}}$$

$$u_0 = \frac{0.1}{1000 \times 3.1 \times 10^{-3}}$$

$$u_0 = 0.03\ \text{m s}^{-1}$$

Ekman's theory states that in the Northern Hemisphere the wind-generated current should be approximately 45° to the right of the wind direction. As we are told the direction of the wind is to the east (i.e. a westerly wind), 45° to the right of this is southeast. This means that the direction of the surface current will be to the southeast.

Question 4.2

In a gyre in the Northern Hemisphere with the surface waters circulating in an anticlockwise direction, there will be a divergence of surface waters. This will depress the surface of the ocean.

Question 4.3

No. The density of water is a function of the temperature and salinity. As each water-mass has a different value of temperature and salinity it will have a different density.

Question 4.4

From Figure 4.12 the density anomaly of a water-mass with temperature approximately 15 °C and a salinity of approximately 35.0 is approximately 26.0 kg m^{-3}.

Question 4.5

Reduced deep-water formation in the North Atlantic could decrease the strength of the deep branch of the oceanic conveyor belt in Figure 4.15. This could have the knock-on effect of reducing the strength of the whole conveyor belt and so less heat would be carried from the Pacific Ocean into the Atlantic Ocean. This could cool the climate of Western Europe — our winters would be much cooler and harsher, and perhaps the climate of Bergen in Figure 1.2 would approach the climate of Halifax.

Question 5.1

If 5.5×10^{12} kg carbon per year are released to the atmosphere and about 35% is absorbed by the oceans, then the ocean absorbs:
$(5.5 \times 10^{12}) \times 0.35 \approx 2 \times 10^{12}$ kg C yr^{-1}.

As the ocean reservoir is approximately:
$38\,000 \times 10^{12}$ kg C, this represents an increase of $\{(2 \times 10^{12})/(38\,000 \times 10^{12})\} \times 100\% = 0.005\%$ per year, and so is apparently not very significant.

Acknowledgements for Topic 5 *Oceans and Climate*

Grateful acknowledgement is made to the following sources for permission to reproduce material in this book:

Cover illustration: SeaWiFS Project; *Figures 1.4, 3.5, 4.5*: Mark Brandon, Open University; *Figure 2.4*: Lamont Doherty Earth Observatory. Map of bathymetry of Antarctic regions modified from an image created by Dr Bruce Huber at Lamont Doherty USA, http://www.ldeo.columbia.edu/physocean/proj_ISW.html; *Figure 3.7*: National Oceanographic and Atmospheric Administration (NOAA); *Figure 3.9, 3.12*: Schlitzer, R., (2000) Electronic Atlas of WOCE Hydrographic and Tracer Data, *Eos Trans. AGU*, 81(5), 45; *Figure 3.10*: Tait, R.V. (1968), *Elements of Marine Ecology*, Butterworth Heinemann; *Figure 3.14*: Apel, J.R. (1987) *Principles of Ocean Physics*, Academic Press Limited. Copyright © 1987 by Academic Press Limited; *Figure 4.2a and b*: National Maritime Museum Picture Library, Greenwich; *Figure 4.3*: Michael Bravo, Department of Geography, University of Cambridge; *Figure 4.9*: Perry, A.H. and Walker, J.M. (1977) *The Ocean-Atmosphere System*, Longman; *Figure 4.11*: Reprinted from Oceanologica Acta, vol. 9, Emery, W. J. and Meinke, J. Copyright © 1986, with permission from Elsevier Science; *Figures 5.1, 5.2*: Albritton, D. L. and Meira Filho, L. G. (2001) *Climate Change 2001*, Intergovernmental Panel on Climate Change.

Every effort has been made to trace all the copyright owners, but if any have been inadvertently overlooked, the publishers will be pleased to make the necessary arrangements at the first opportunity.

TOPIC 6

WATER QUALITY

Colin Chapman

Introduction

1

1.1 What do we mean by water quality?

In parts of the world that suffer from severe drought or devastating floods, populations are likely to have attitudes towards water that are concerned with quantity as well as quality. Humans can survive 50–60 days without food but only 5–10 days without water. In the UK we have a love–hate relationship with water: we drink it, wash with it, swim in it, often waste it, worry when our gardens are dry, complain when it rains but generally assume that its quality is 'fit for purpose'.

Figure 1.1 The Matterhorn, Switzerland illustrating the physical states of water. (Clouds are actually droplets of water rather than water in the gas phase.)

Let us consider what is generally meant by **water quality** before we take a detailed look at the chemistry and other scientific factors. Accepting that the wording is a shorthand way of writing 'quality of water', let us look in an English dictionary, that is, one published for general readers, not for people who may be studying science, nor for anyone with a particular specialist background. We find that the relevant definition of *quality* is 'relative nature, kind or character'; we also find that *water* is defined as 'a colourless, transparent, tasteless, odourless, compound of oxygen and hydrogen in a liquid state, convertible by heat into steam and by cold into ice' (Figure 1.1).

You know about the conversion processes of water into steam and into ice and about the temperatures at which these conversions take place. You may recall that an input of energy is required to bring about the change in state from ice to liquid water and from liquid water to steam (or gaseous water). If you have travelled by road in wintry conditions, you will be aware that an impurity reduces the freezing temperature of water (or the melting temperature of ice). Salt is scattered on icy roads and footpaths by highway authorities (Figure 1.2).

- ❍ What happens when salt is scattered on an icy road or footpath?
- ● Salt reduces the melting temperature of ice, which then melts. However, if the temperature is very low (less than about −20 °C), this treatment is not effective.

Pressure affects the boiling temperature of water: lowering the pressure lowers the boiling temperature. There are changes in volume that occur when there is a change of state (from solid to liquid and then from liquid to vapour). The simple, almost dismissive, dictionary wording 'compound of oxygen and hydrogen' to describe water says nothing about the forces that hold the atoms together in the molecule and nothing about the forces between individual molecules.

Figure 1.2 A road that has been cleared of ice and snow.

You may not agree with the dictionary that water is colourless, transparent, tasteless and odourless. Water is none of these in an absolute sense. Perhaps you have come across water that is strongly coloured, or at least tinged with some colour, or is cloudy or murky,

and has a particular taste and a specific smell that you find attractive or perhaps repulsive. Even very pure water is faintly blue in large quantities. Obviously, this is where the definition of quality as being 'relative character' comes in. But 'character' is also a rather vague term when applied to water, for some individuals immediately think of chemical or biological quality while others think of physical quality. In practice, all three aspects — chemical, biological and physical — are of interest to authorities and organizations that have responsibility for water purity. Some water users (Figure 1.3) are more concerned with chemical quality such as its hardness (see Section 3.2), some with biological quality, such as its microbe content, and others with the physical quality of water such as its temperature or even its refractive index or surface tension.

Water may be highly coloured or turbid, or be quite clear (Figure 1.4). We might be tempted to believe that a colourless water has a *higher* quality than a coloured water and that the colouring is an undesirable impurity. However, it would be more correct to say that the colourless water has a *different* quality. The colour, transparency, taste, odour, or lack of these, could be an advantage to some users but a disadvantage to others. In other words, our interpretation of water quality depends very much on the purposes for which the water is to be used, by whom or in which process, and on many other factors that we shall meet in the following sections. The analytical techniques used to discover what, besides water molecules, is in a particular water sample, and the factors that affect its quality, also depend on the interests of a specific user of the water. It is important to realize that water of a high purity is not necessarily water of a high quality, which is a non-absolute property.

Figure 1.3 Some of the uses of, and demands placed upon, water: transport, leisure, personal and clothes washing.

(a)

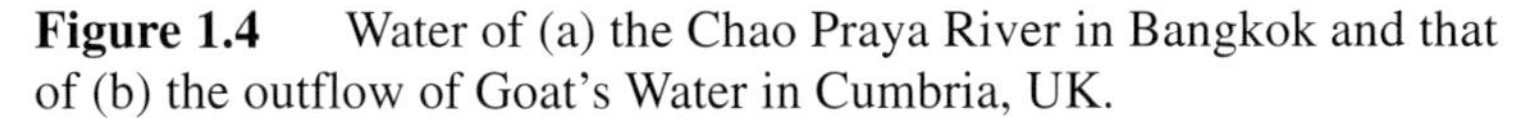

Figure 1.4 Water of (a) the Chao Praya River in Bangkok and that of (b) the outflow of Goat's Water in Cumbria, UK.

(b)

Our dictionary then goes on to mention many categories of water, including hot and cold water, salt and fresh or sweet water, hard and soft water, aerated water, saline water, thermal water and 'strong water'. There are other categories or descriptions of water; some descriptions may be associated with the circumstances in which the water is used and these descriptions may be associated with its quality: whether the water is potable (fit for drinking) for example.

1.2 The hydrological cycle and water quality

You know that much of the water in this world of ours is cycled naturally on time-scales ranging from hours to thousands of years. The bulk of the world's water is in the salty oceans, which cover more than 70% of the surface of the Earth and have a volume of about $1.34 \times 10^9\,\text{km}^3$. (Saltwater can be defined as water containing more than about four grams of dissolved salts per litre.) Freshwater is found frozen in the polar ice caps and mountain glaciers and as a liquid in lakes, rivers and groundwater. To complicate things, there are also saltwater lakes and there is a liquid water equivalent of $1.3 \times 10^4\,\text{km}^3$ in the atmosphere. In total, the Earth has about $1.39 \times 10^9\,\text{km}^3$ of water of which $3.7 \times 10^7\,\text{km}^3$ is freshwater. The largest body of liquid freshwater is Lake Baikal in southeastern Siberia, with a volume of $2.3 \times 10^4\,\text{km}^3$. It is the deepest lake in the world with a mean depth of 740 m and a maximum depth of 1620 m. The Caspian Sea contains saltwater (as the inclusion of 'sea' in its name implies) and has the largest surface area of all the lakes in the world. Its area of $3.71 \times 10^5\,\text{km}^2$ is significantly greater than that of the whole of Great Britain ($3.15 \times 10^5\,\text{km}^2$). (See Figure 1.5.)

The part of the hydrological cycle where water evaporates from the oceans into the atmosphere represents **desalination** on a massive scale. Water also evaporates from inland lakes and rivers and from vegetation and soil. On being cooled, this

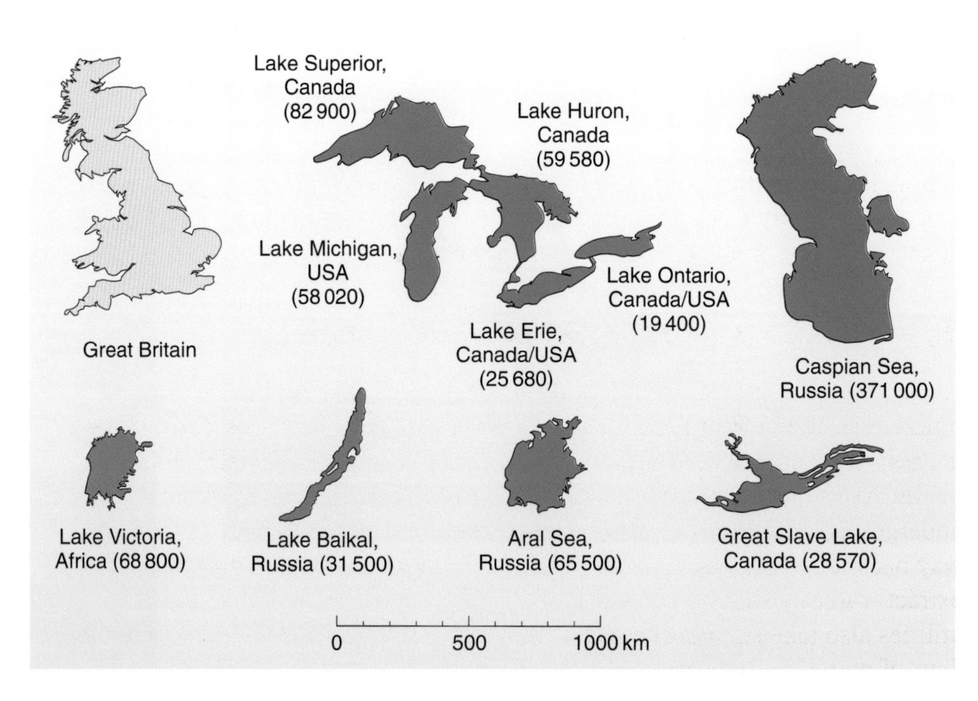

Figure 1.5 Relative areas of the major lakes of the world. Great Britain is also shown on the same scale. (Areas are expressed as km^2.)

vapour condenses as relatively pure water and returns to the seas or the Earth land as a precipitation of rain, hail, sleet or snow. As rain or other precipitation falls through the air, its quality changes as it dissolves or entraps gases, salts, organic compounds and small items of particulate matter.

Precipitation falling on land runs across (by overland flow), or soaks through (by infiltration), the rocks and soil, picking up mineral salts and other materials on its travels. Some of these materials are carried in suspension in the water, others dissolve as solutes depending on their solubility in the water. Solubility itself depends on the particular water sample in terms of its pH, its temperature and other solutes already dissolved in the water. However, some of the precipitation does not fall directly on the ground but is intercepted by vegetation or urban development (e.g. roads, car parks, buildings).

Water flowing over or penetrating the soil can take on a quite different quality to that of the initial precipitation. It finds its way back to the oceans by streamflow (passing into streams, along rivers and through lakes) or by groundwater flow (through the ground). Streamflow may get water to an ocean in a matter of days or weeks, but it is important to appreciate that it may take decades or more for water to reach the oceans by groundwater routes. Sometimes streamflow is diverted through constructed canals and along aqueducts into reservoirs or cisterns. (Small reservoirs in some areas are called cisterns.) Huge quantities of inland water are evaporated directly back to the atmosphere from fields and hillsides, from the surfaces of rivers, lakes and ponds and from surface cover, both artificial and natural. The rate of evaporation depends upon humidity, temperature, wind, albedo and soil texture and depth.

In some places, slow-moving rivers and streams, having passed over or through soil, picking up materials during their passage, find their way into a land basin but do not flow out again. Water is lost by evaporation but this is replenished by the inflow carrying its load of solutes and suspended materials. Evaporation removes only water vapour and not the materials suspended or dissolved in the water. These materials become concentrated in the lake that eventually turns into an inland sea, for example the Caspian Sea in Russia, the Great Salt Lake in Utah USA, or the Dead Sea in the Middle East.

Considerable quantities of water are taken from the soil by vegetation, although much of this water is subsequently returned to the atmosphere by transpiration routes. All organisms from the simplest to the most complex take in and discharge water across their cell membranes.

1.3 Water users and water quality

All organisms consume considerable quantities of water either directly as a liquid or as an inherent constituent of a wide variety of nutrients. Freshwater is necessary for human consumption so we turn to water sources derived from the relatively pure precipitation. Groundwater may be extracted from underground sources by water companies before it has a chance to wend its way back to the oceans. Besides tapping into these underground sources to extract water by sinking boreholes and by digging wells, individuals and utilities also take streamflow water from lakes, streams and rivers. Before the days of domestic, industrial, or even agricultural pollution, these processes presented little problem. However, discharges of effluents into the environment, particularly during the past 500 years, have meant that some precautions or treatments are necessary before most water sources (both underground and surface) can be safely used for human consumption.

Large volumes of water are used by industrial and domestic consumers. Up to 10 litres of water can be used in the production of one litre of beer, 200 litres of water in the production of one newspaper, and 50 000 litres of water in the production of one car (Figure 1.6). In the western world, this water is provided by water companies. They extract the water from the environment and treat it to drinking water standards before distribution to their customers. Some industries, such as a pharmaceutical, semiconductor or microelectronics components manufacturer, or a power plant using steam at very high temperatures and pressures, require water that has far fewer, and lower concentrations of, solutes than drinking water. The water then must be further purified and treated.

In many cases, however, consumers do not require water of potable quality because they are never going to drink it or use it in manufacture. Local authorities use water for street-cleaning services and roadside fire hydrants. Domestic consumers use water to flush toilets, irrigate gardens and wash clothes and cars, for example, and for none of these activities is drinking-quality water necessary. Many industrial consumers could equally use water of a lower than potable quality for a large proportion of their needs. And yet, all over the world, a great deal of money is spent in what could be regarded as 'over-purifying' or 'over-treating' water, just in case a consumer may wish to use it for drinking or preparing food. In some places, there have been attempts to address this situation: in San Francisco, for example, there is a separate saltwater supply for fire-fighting use; whilst in Coalinga, California, homes have dual water supplies, with seawater for sanitary use and potable water for drinking and cooking. Generally, in Europe, it has been considered economically prohibitive, and also a potential health hazard, to install dual supply systems.

Figure 1.6 Some products requiring use of water in their manufacture: vehicles, beer, paint, detergents.

Whilst making use of the water piped onto their premises, users put all manner of waste products into it before returning it to the environment. Even the Romans had flushing toilets. Domestic users take water for bathing their bodies and washing their clothes, returning it with a different quality, containing not only 'dirt' but also other impurities such as partly-used soaps and detergents used in the washing.

Some industries, including breweries and other drink manufacturers, and producers of low-fat spreads and other food, use water as an essential ingredient of their final product and, having treated it suitably, effectively sell this water on to their customers. These enterprises also use large volumes of water to rinse bottles, clean vats, wash floors and cool a variety of processes. Fruit and vegetable canning and bottling factories also not only use water in their washing and preparation processes but they sell the included water, again appropriately treated, in the cans and containers of their products. The dairy industry similarly uses water for washing equipment and products and for cooling, but additionally uses water in reconstituting some products; between 4.5 and 8 litres of water are needed altogether to prepare one litre of milk. Sugar beet and sugar cane mills all use water to prepare and purify sugar juice, while sugar refineries use water in decolorizing and refining processes. Other industries, such as paper and board manufacturers, wool and cotton textile producers, dyeworks, oil transporters and refiners, automobile and aeronautical industries, and steel and other metal fabricators with blast furnaces, coking and descaling plants and rolling mills, all use water as a vital agent or ingredient, but then remove that water during processing, before selling the final manufactured or processed items to their customers.

Some users take water for cooling an industrial process and then return it to the environment at a higher temperature; the only alteration in water quality from these users is temperature change. However, temperature affects the concentrations of solid and gaseous solutes in water. Warmer water may be regarded as being of a higher quality by some but of a lower quality by others. In cases where the water's temperature is likely to become unacceptably high, it is cooled in large towers before being reused or returned to the environment. For those industrial processes that use water for quenching, as in the production of cast iron, or as a coolant for cutting tools, the water does not merely warm up as the components are cooled, but it also becomes contaminated.

Other users merely take advantage of the energy in naturally-flowing water to drive watermills, so producing mechanical power, or turbines to produce electrical power. The Cotswolds woollen industry in the 18th century developed with small watermills built on streams throughout the valleys in that part of England (Figure 1.7). Most of the electricity in Ontario, Canada comes from hydroelectric power stations, where the 'fuel' costs are negligible compared with those at a fossil fuel-fired power plant. Such users subsequently return the water to the environment without affecting its quality.

Some hydraulic systems, such as metal presses in the motor manufacturing industry, use water as the power-transmitting medium. In this case, however, the water pressure is derived artificially from pumps, not from nature's resources, which water turbines make use of. A small piston in a small chamber is connected to a large chamber covered by a large piston. By applying a small force to the small chamber, and so moving the small piston a considerable distance, the large piston is pushed along in the large chamber, albeit through a short distance. Water quality in hydraulic systems has to be such that there is no corrosion of components and no other particulate matter that could interfere with the smooth travel of a piston within its cylinder. Water under pressure is also used in the petroleum industry in the secondary recovery of oil and in the metal industry to descale steels.

Yet other users take nothing from the water at all and simply use it as a highway. On occasions, commodities, such as logs, are moved by river water simply by rafting them together and floating them downstream, a familiar sight in the forestry industry in North and South America. Canals, rivers and the oceans are used to transport people and goods from place to place, as well as for recreational purposes. The water quality required in all these cases is normally that it be free of major obstructions. Archimides' principle and buoyancy and displacement volumes may determine whether objects float or sink, but one essential quality of water, its density, is a vital factor. Whilst the above carriers are taking nothing *from* the water, they occasionally – by mistake or deliberately – discharge waste products or even their cargoes *into* the water and so change its quality adversely.

Figure 1.7 A watermill in the Cotswolds, UK.

Another area where water is used as a means of transport is in central heating systems, where heat is simply transferred from a boiler by pumping water through pipes to a series of radiators in rooms around the home or workplace. Water is ideal for this purpose because it is cheap, has good heat retention properties, has a high specific heat compared with other liquids, and its relatively low viscosity and density enable it to be readily pumped around a heating system without too much power consumption. Nothing solid is being transported in the water, only energy in the form of heat. However, to protect the metal pipework and radiators, it is common practice to add chemicals to the circulating water to inhibit or minimize corrosion.

The nuclear power industry makes use of the heat transfer properties of water and its ability to slow down and absorb ionizing radiation. Spent fuel elements (the sealed containers enclosing the nuclear fuel) from a nuclear reactor are hot and radioactive. They are stored under water in special cooling ponds where they can be observed during the minimum period of 100 days that is necessary for the short-lived radioisotopes to decay safely. The water acts as both a transparent radiation shield and a coolant. The quality of water in a fuel element storage pond is maintained by circulating it through coolers and a chemical treatment plant to remove corrosion products or radioactive substances from any leaking fuel elements.

The transparency of water is a function of the wavelength of the radiation passing through it. Water is relatively transparent to most visible radiation but there is significant absorption in the infrared, with a small tail extending into the red region of the spectrum, resulting in the pale blue colour of large quantities of water. Visible light is transmitted well but infrared light can hardly penetrate water.

We should not overlook other uses of water, for therapeutic and recreational purposes, for example. For centuries, towns and resorts have been built around water sources with a special quality — they are naturally rich in certain minerals considered beneficial for health either by drinking the water or, more commonly, bathing in it. At many of these centres, the water temperature is also somewhat above the ambient value. The original inland watering place is the town of Spa in Belgium, but many similar 'spa towns' in Europe were created by the Romans. Some were popular in earlier times, while others were developed commercially in later eras. In England, places such as Bath, Buxton, Cheltenham, Droitwich, Leamington and Harrogate drew guests from vast distances to 'take the waters'. St Moritz in Switzerland, Vichy in France and Baden-Baden in Germany offered similar facilities. These spas depend for their survival on impure water whilst claiming positive outcomes from the impurities.

Public pools for swimming and water polo, and ice rinks for skating, curling and ice hockey, do not demand vast amounts of water on a continuous basis. However, the water for these has to conform to quality standards for specific needs. Swimming pool water is constantly being polluted by bathers with skin, throat and faecal bacteria, saliva, hair, nitrogenous matter from sweat and urine, and with cosmetics, body oils, even sticking plasters and hair clips. All these materials must be constantly removed to maintain the water at a quality that is acceptable for all users. To minimize the transfer of infections and the growth of fungi and algae, swimming pool water also has to be continuously disinfected. The major quality of water for an ice rink is a low temperature rather than chemical purity. However, too many impurities in the water will reduce its freezing point and thus lead to higher operating costs in maintaining a frozen surface for the skaters.

Ice for use as a cooling medium is subject, of course, to yet other considerations, again depending on the exact application. When ice is in direct contact with food or drink for human consumption, its quality nowadays must meet the same requirements as for those foodstuffs, but if it is isolated from them by a suitable barrier, its quality may be permitted to differ. Until about 70 years ago, large quantities of natural ice were harvested in places such as Norway and stored in ice-houses for distribution when needed, especially to Britain.

Ice-houses in a quite different sense of the term, but more correctly snow-houses, are known in Europe as igloos. These are used by the Inuit peoples of Alaska, northern Canada and Greenland in the winter hunting season as temporary dwellings, though they are normally up to only about 3 metres in diameter and 3.5 metres high. Such a snow-house, called *igluigaq* in the language of the Inuit, is constructed of blocks of hard-packed snow, fitted together in a spiral with the top blocks leaning in to form a dome. A window of clear ice is sometimes inserted in the side (Figure 1.8). The quality of the ice or snow that is of primary importance in this situation is its temperature.

We should not forget that water is used by many organisms as a permanent home — it is their natural habitat. Fish and aquatic creatures, aquatic plants and algae depend entirely on an aqueous environment for survival; when exposed to air for extended periods, they simply dehydrate and die. In most cases water quality for them is primarily concerned with temperature and the concentration of dissolved gases, coupled with an absence of toxic materials. At normal temperatures, only about 30 cm^3 (0.030 litres) of oxygen will dissolve in one litre of water. Nevertheless this amount is normally sufficient to support aquatic life. But if, through temperature changes or the addition of reducing agents and oxygen scavengers, oxygen levels fall much below this level, life in the river or lake will change.

Figure 1.8 Ice-house built from blocks of compacted snow.

Water is also used to provide structural support in both plant and animal tissue. Seedlings, non-woody plants and soft parts such as leaves and flowers are quite limp when their tissues have lost water. Most vegetables, and fruit in particular, depend on the pressure of water within their cells and tissues to keep them firm. Many tissues in birds and fish, but not in insects which have a rigid external skeleton, also depend on water to maintain their shape and firmness.

The anhydrous, usually powdered and amorphous, forms of some substances swell and become rigid in crystalline shapes when water is added. Some anhydrous materials are hydrated with the addition of water of crystallization. A specific number of water molecules are needed for a particular crystalline compound. For example, sodium sulfate and sodium carbonate require ten water molecules each for each formula unit, and magnesium sulfate requires seven. Some substances, such as sodium chloride and common salts of potassium and ammonium, crystallize in their anhydrous form, in other words, without any water of crystallization. The quality of the water in terms of impurities is important when crystals are grown naturally because any impurities will cause the crystals to form irregularly, as well as providing contaminants.

1.4 Humans and water quality

The human body is made up of about 65% by mass of water circulated through our body systems as the carrier or solvent for very many essential chemicals. For an average-sized adult, this 65% amounts to around 55 litres of water. Of this volume about 40 litres are intracellular (inside the cells) while the remaining 15 litres are extracellular in various body fluids. We either drink it, or obtain it indirectly through the water held in plants and animals that we eat. Green vegetables and fruit contain from 78–97% water; fish about 80% water, and meat about 72% water. (In comparison, marine organisms, including jellyfish and algae can contain over 98% water, whereas bacteria can have as little as 50% water in certain conditions, enabling them to survive when other life forms are vulnerable.)

The water in our bodies is not distributed evenly; bones contain from 22–34% water while nerve tissue has from 82–94% water. Our blood consists of about

50% water, the remainder consisting of red and white blood cells and platelets and plasma substances. The blood system transports oxygen and nutrients to various cells and removes carbon dioxide and other waste products for detoxification and elimination. Toxic ammonia, for example, a waste product from the digestion of proteins, is converted by enzymes in the liver into the less poisonous urea, $CO(NH_2)_2$, which is excreted from the body in urine. This conversion is extremely efficient as the urea is formed from the unwanted ammonia with carbon dioxide, another of the body's waste products:

$$2NH_3 + CO_2 = CO(NH_2)_2 + H_2O$$

Water is essential for our bodies to process our food. We eliminate water (with other waste products) in our breath (14%), sweat (19%), urine (59%) and faeces (8%). The body's water balance is maintained by our kidneys which, on an average day, filter out some 150 litres of water from our blood but reabsorb over 99% of the water back into the blood, expelling the remainder in urine. Dialysis is an artificial method of body fluid filtration, which is utilized medically in cases of kidney failure and is also the principle used in some industrial water filtration processes. Our bodies need a throughput of about 2.7 litres of water in one form or another every day, the quantity depending on the type of work we do, from manual labour to sedentary activities, and also on where we live, from Saharan Africa to chilly Iceland. Our drinking water must not contain too many pathogenic microbes, although our stomachs can cope with other impurities such as the hardness salts, calcium and magnesium carbonates and hydrogen carbonate. In contrast, identical concentrations of hardness salts can cause untold damage in a commercial or domestic hot water system which is unaffected by the microbes that might harm us.

This means that water quality that is perfectly acceptable for one purpose or user can be totally unacceptable for another. We may need to remove hardness salts (see Section 3.2), for example, from water that is kept in a closed loop and circulated around a heating system, but we may need to add something to water to kill microbes before it is fit for us to drink.

Our demands for water and our requirements for water quality vary enormously. We should distinguish carefully between our *wants* and our *needs*. In the developing world, the combined industrial, agricultural and domestic annual water consumption amounts to an average of 100 m^3 per person, while in the USA the figure is 1500 m^3. The world's use of water averages around 250 m^3 per person per year, although this is continually rising as demands for higher standards of living increase.

It is more than likely in today's world that a water source will be far from the intended point of domestic or industrial usage. For agricultural users, water for irrigation has to be taken to the areas where crops are growing. The quality of irrigation water may also need to vary from that required by either the domestic or the industrial user. Thus water is conveyed by gravity or pumps through pipes or transported by some other means from the source, possibly to a local service storage facility, and then distributed to the end-users — the consumers and customers. The distribution system itself may become a source of contamination by introducing impurities. Decisions regarding whether and where to install purification plants hinge on evaluating such situations.

There is also the occasional reverse of conveying water *to* a location, when flood waters have to be pumped *away from* an area to enable animals to graze, crops to be grown, people to return to their homes, and everyday life to return to normal. Normally those responsible for reducing floods and similar inundations are mostly concerned about the quantity, rather than quality of such water — unless it is carrying dead and rotting carcasses, raw sewage, or vast quantities of unwelcome sediment, and so poses a health hazard.

1.5 Summary of Section 1

1 There is no absolute definition of water quality. The quality of water is a function of the extent to which it is 'fit for purpose'.

2 Water features in almost every form of natural process and, directly or indirectly, in most forms of human activities.

2 Properties of water

2.1 Solutions, solubility and equilibria

The omnipresence of water is as a result of some rather special properties. Water molecules comprise two hydrogen atoms bonded to an oxygen atom. These molecules are relatively stable in that they are not decomposed under the normal conditions on Earth. An input of energy of 242 kJ mol^{-1} is required to bring about the decomposition:

$$H_2O(l) = H_2(g) + \frac{1}{2}O_2(g)$$

In other words, it takes as much energy to decompose 18.0 g water (one mole) into hydrogen and oxygen gas as it does to heat *forty* times this amount of water from 20 °C to 100 °C. Not only are the molecules themselves rather robust but hydrogen bonding results in significant attractive forces between the molecules. As a result, water has a higher freezing and boiling temperature than one might expect for a molecule of similar size. Conveniently, the liquid range of water (0 °C to 100 °C) spans the temperature range of most of the surface of the Earth. If you take some sodium chloride (common salt) and stir it into a beaker of cold rainwater the salt will dissolve to form a solution that resembles seawater (although true seawater contains very many other solutes such as calcium and magnesium carbonates, chlorides and sulfates). If you continue to add more sodium chloride, this will also dissolve until a point is reached when any further sodium chloride added remains undissolved. The salt solution is now saturated with respect to sodium chloride.

You should be aware that industrial water chemists do not always use chemical terminology with the same precision as do university lecturers or research students. You will frequently find that the term 'salt' is applied to almost any ionic solute in water and sodium chloride (NaCl) is also given the name 'salt' or 'common salt', even though it is just one example of a salt. Other ionic compounds such as calcium sulfate and magnesium chloride are also frequently termed 'salts' in the context of water chemistry.

Water is a polar solvent, which means it has small negative and positive charges in the molecule. Polar solvents tend to dissolve ionic compounds, which is consistent with sodium chloride having good solubility in water. Molecules containing the $-OH$ (alcohols and sugars), $-SH$ and $-NH_2$ groups tend to have some solubility in water and are termed **hydrophilic** (water-loving). **Hydrophobic** (water-hating) groups include $-CH_3$, $-CH_2-$ and $-C_6H_5$ and impart a lack of solubility in water. Non-polar solvents, such as tetrachloromethane (CCl_4), liquid hydrocarbons and oils, are generally poor at dissolving ionic compounds but are good at dissolving non-polar compounds.

Solutes do not have to be solids. Alcohol, a liquid, is also soluble in water, and so alcohol is a liquid solute. Because water acts as a solvent for alcohol, manufacturers can easily produce alcoholic drinks such as vodka, gin and whisky with differing strengths, or proofs, simply by adding differing amounts of water. Gases such as oxygen and carbon dioxide, can equally well be solutes. Fizzy drinks, such as soda water, are good examples of solutions of carbon dioxide (the solute) in water (the solvent). Fish and other aquatic life depend on the oxygen in air dissolved in freshwater and seawater. Air is itself a solution of various gases in each other, so in this case the gases are both solutes and solvents.

The solubility of a solute in a solvent, at a given temperature and pressure, is defined as the maximum mass of the solute in grams that will dissolve, at that temperature and pressure, in 100 g of the solvent (in the presence of an excess of undissolved solute). A substance having solubility at ambient temperatures of less than 1 g per 100 g of water is normally regarded as being insoluble in water. Generally the pressure on a solution does not affect the solubility of solid solutes very much, although at extremely high pressures, such as can occur in steam raising units or boilers at power stations, solubilities are often higher.

At the point where the solvent can only just accept no more solute into solution, there is equilibrium between the solute and solvent. In general, equilibrium is reached when the rate at which a reaction moves in one direction is equal to the rate of the reversible reaction. Equilibrium systems follow Le Chatelier's principle that 'for a system in equilibrium, if any change is made to the conditions, the equilibrium will alter so as to oppose the change'. Changes in conditions include temperature, concentration and pressure (in the case of gases).

For many (but not all) ionic compounds, solubility is greater at higher temperatures (Figure 2.1). Sodium chloride, however, is not at all typical as its solubility is hardly affected by temperature. However, the solubility of sodium nitrate is twice as high at 75 °C as it is at 0 °C, and that of potassium nitrate is twice as high at 50 °C as at 25 °C. If a saturated solution is cooled rapidly, a supersaturated solution may be created containing an amount of solid in excess of its true solubility.

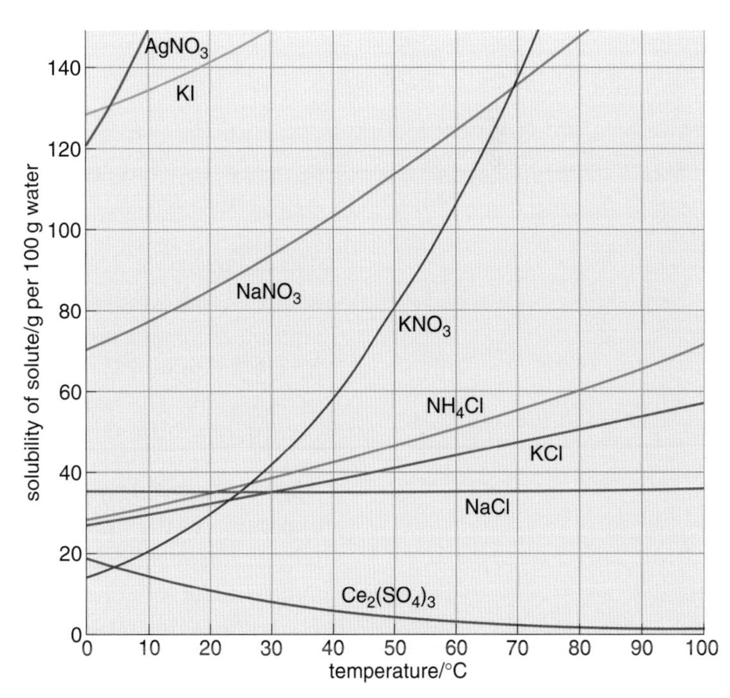

Figure 2.1 Solubility of some ionic compounds in water over a range of temperatures: Silver nitrate, $AgNO_3$; potassium iodide, KI; sodium nitrate, $NaNO_3$; potassium nitrate, KNO_3; ammonium chloride, NH_4Cl; potassium chloride, KCl; sodium chloride, NaCl; cerium(III) sulfate, $Ce_2(SO_4)_3$.

Question 2.1

What would you expect to happen if an aqueous solution of volume 1.5 litres containing sodium nitrate ($NaNO_3$) at a concentration of 110 g sodium nitrate per 100 g water at 60 °C, were cooled to a temperature of 30 °C?

Some compounds are *less* soluble in hot water than in cold water. Examples of these are calcium hydroxide and calcium sulfate. Calcium sulfate occurs very commonly in river and seawaters, and because it is harmless to the human body it is rarely removed from potable supplies. In hot water systems, there is a possibility of precipitation because of its unusual solubility characteristics. The solubilities of some solutes found in both river water and seawater are shown in Table 2.1 and the average concentrations of ions in water from natural sources in Table 2.2. It is worth noting that, in general, gases are *less* soluble with increasing temperature.

Table 2.1 Solubilities of some magnesium and calcium salts in water at 25 °C and 100 °C.

Solute	Solubility at 25 °C /g per 100 g water	Solubility at 100 °C /g per 100 g water
calcium carbonate	0.0013	0.0015
calcium sulfate	200	180
calcium chloride	68	159
calcium nitrate	120	280
magnesium carbonate	0.011	0.018
magnesium hydroxide	0.0011	0.0011
magnesium sulfate	33	74
magnesium chloride	52	65
magnesium nitrate	75	very high

Table 2.2 Average concentrations of ions in water from natural sources.

Ion	Concentration/mg l^{-1}				
	seawater	world average, freshwater	Lake Ontario, Canada	Lake Windermere, UK	Lake Baikal, Russia
Na^+	10 810	8	12.2	3.5	—
Ca^{2+}	410	38	43	5.7	15
Mg^{2+}	1300	5	6.4	0.6	4.2
HCO_3^-	140	105	115	9.7	60
Cl^-	19 440	8	27	6.6	1.8
SO_4^{2-}	2710	18	27	6.9	4.9

The solubility of a solute in a solvent will also affect the solubility of another solute in that same solvent. In other words, water that has become saturated with sodium chloride cannot dissolve as much calcium carbonate as can a sample of newly fallen rainwater. Nutrients in body cells also follow the same principles of solubility. The nutrients are the solutes and the water, acting as a carrier, is the solvent.

The solubilities in water of the main atmospheric gases are shown in Table 2.3, expressed as the number of litres of gas that dissolve in a litre of water at the temperatures stated. Remember that the Earth's atmosphere contains approximately 80% nitrogen, 20% oxygen and around 0.03% carbon dioxide. Each gas in this table is assumed to be at a pressure of 10^5 Pa (1 bar) when in equilibrium with water. You can see, as mentioned above, that the solubilities of these gases *decrease* as the temperature rises.

Table 2.3 The variation of gas solubility in water with temperature.

Gas	Solubility/litres of gas per litre of water at 10^5 Pa			
	0 °C	15 °C	30 °C	60 °C
carbon dioxide	1.68	0.99	0.65	0.36
oxygen	0.023	0.017	0.013	0.010
nitrogen	0.047	0.033	0.026	0.019

The amount of dissolved carbon dioxide in water also depends on the pH. Below pH 5, carbon dioxide is essentially present as a dissolved gas; from pH 5 to pH 9 some of the carbon dioxide is present as the hydrogen carbonate anion, HCO_3^- and above pH 9 there is significant carbonate anion CO_3^{2-}. Acid rain (rain with a reduced pH) is able to dissolve more calcium carbonate from soils than can rain with a higher pH.

Water chemists often use the term **total dissolved solids (TDS)** to identify the approximate quantity of all substances dissolved in a particular water sample. Measuring TDS gives an overview of the quality of that water with respect to dissolved solids. The relative amounts of TDS in the three broad categories of typical British inland waters are shown in Table 2.4.

Table 2.4 Range of values for concentrations of dissolved solids in inland waters. (Values given in parts per million by mass, ppmm.)

Water category	Range of total dissolved solids/ppmm
moorland rivers and lakes	40–80
springs and lowland rivers	150–300
deep well waters	250–450

If two electrodes are immersed in water and a current is passed between them, the electrical resistance will be lower in water with a higher TDS value because many of the dissolved solids in water are ionic solutes with electrical conducting properties. Conversely, the lower the TDS value of a water sample, the higher the electrical resistance. Measuring the electrical resistance of a water sample is a cheap and rapid method of assessing the TDS value, although it has to be noted that resistance changes with temperature. A standard test cell comprises a water container with two electrodes each of area 1 cm^2 and separated by a distance of 1 cm so that resistivity (in units of ohms centimetres) can be measured. In practice, the test cells are calibrated to indicate the reciprocal of electrical resistivity, electrical conductivity, with units of Siemens per centimetre (sometimes termed mho — ohm written backwards) or its subunit, microsiemens per centimetre ($\mu S\ cm^{-1}$). A resistivity of 100 000 ohms centimetres equates to a conductivity of $10\ \mu S\ cm^{-1}$. The variation in conductivity is shown in Figure 2.2.

Deep well waters often have high conductivity values; moorland rivers and lakes usually have lower conductivities; distilled water even lower and seawater very much higher. Conductivity increases as water temperature increases.

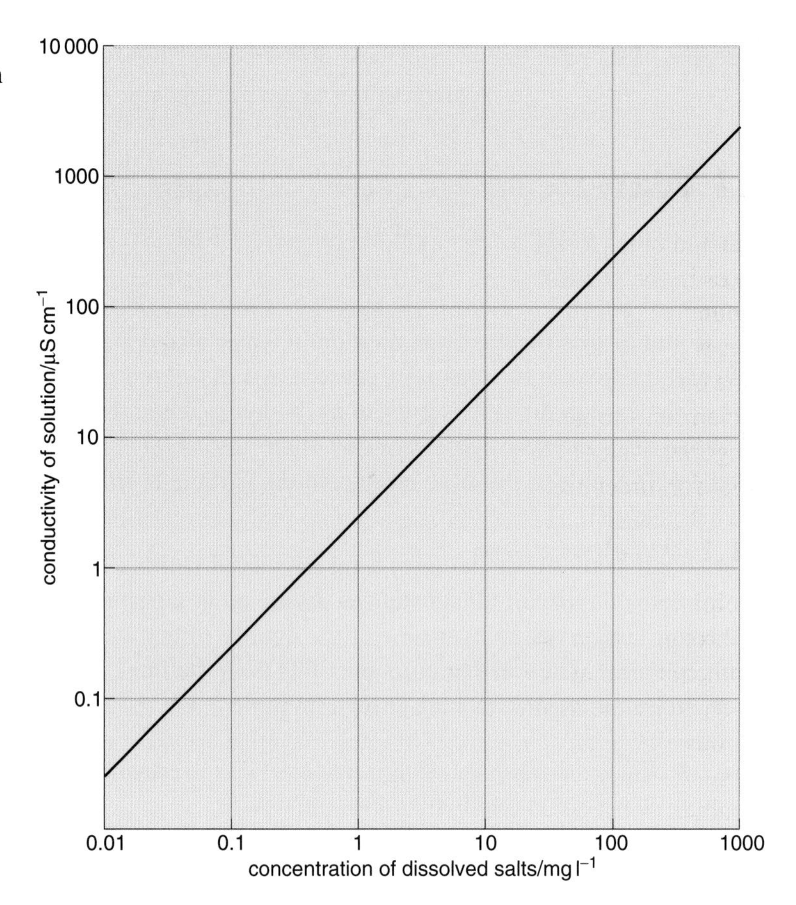

Figure 2.2 Variation in conductivity with concentration for solutions of dissolved salts (such as sodium chloride) in water.

Water is an extremely good solvent for solids (especially ionic solids), liquids and gases. While this may be advantageous for all the reasons that we have described, particularly for the ability to carry nutrients into animal and plant cells, this property is the sole reason why it is so difficult for us to find or obtain water that is chemically pure.

2.2 Summary of Section 2

1 Solubility of ionic solids in water generally increases with increasing temperature but there are several exceptions to this.

2 Solubility of gases in water decreases with increasing temperature.

3 Conductivity can be used to estimate the total amount of dissolved ionic substances in water.

Categories of water

3

3.1 Water categorized by source

Rain and other forms of precipitation eventually find their way to the sea. Arid places in the world receive very little precipitation, other places receive greater amounts depending largely on the latitude, proximity of mountains, season of the year, winds and temperatures. Some precipitation falls directly on the oceans, some on frozen ice caps and mountainous regions and becomes trapped in glaciers, and some falls on land of various qualities. One convenient way to categorize water in terms of its quality is by the source from which it is drawn, either for direct use, or for use after some processing.

3.1.1 Precipitation

Precipitation can be rain, sleet, hail or snow, depending on the temperature of the air through which the condensed water vapour falls to the ground. Small amounts of precipitation are contributed by dew and frost. Rain quality, particularly, is governed by the quality of the air, which may contain domestic or industrial gaseous effluents in addition to the atmospheric gases, nitrogen, oxygen and carbon dioxide. Near the sea, particularly on windy days, salt spray may become entrained in the rain, contributing some corrosive qualities. During prolonged dry periods, the atmosphere may contain relatively high concentrations of dust and pollen so that any precipitation would entrap these, and also other airborne substances such as volcanic ash.

As noted earlier in the course, rainwater is always slightly acidic (i.e. it has a $pH<7$) because of the dissolution of atmospheric CO_2. But in polluted environments, the pH of rain (and other forms of precipitation) is often lower still. The main culprits are oxides of nitrogen and sulfur generated during the combustion of fossil fuels — in domestic heating systems, cars, trains, power plants and many industrial processes via oxidation pathways within the atmosphere (explored in more detail in 'Acid Rain'), these gases are converted to the readily soluble nitric and sulfuric acids, and washed out in rain. Particulate matter and other gases (or the products of their reactions within the atmosphere) released by human activities can also be captured by precipitation — affecting its 'chemical' quality.

3.1.2 Inland water

Some precipitation water runs across hard rocks like granites, which are often covered with mosses and lichens, picking up organic matter, possibly running through heather roots and peaty areas into moorland lakes and rivers. Some falls on soft limestone rocks or chalky or alkaline soils, possibly penetrating the topsoil and emerging later into springs and lowland rivers. Some percolates through the soil until it meets deep-lying impervious rock strata and then emerges very much later. On top of the impervious layers, the water may be trapped in absorbent layers (aquifers) until artificially extracted by the sinking of deep wells and boreholes. During periods of prolonged sunshine, soils that are normally absorbent may be surface-baked relatively hard by the Sun; the result is that any immediate heavy precipitation, say from cloudbursts during thunderstorms, can become an overland flow. This means that the rain runs directly off the land into watercourses, washing material from the surface of the soil with it, rather than following its usual path of infiltration. A similar effect, but called saturated overland flow, occurs when the subsurface is saturated with water after prolonged rainfall. Although there are many variations and combinations of the routes for water flow, British inland water supplies can be conveniently categorized by these routes into three regions of (a) surface waters draining from peaty areas (often acidic), such as moorland lakes and rivers; (b) surface waters that drain alkaline soils, flowing to springs and lowland rivers; and (c) deep well waters.

Analysis of the inland waters from the three sources given above shows interesting differences. Category (a) normally contains essentially no hardness-inducing compounds, such as calcium and magnesium hydrogen carbonates, carbonates and sulfates (see Section 3.2). The water is regarded as soft, although it is likely to contain organic matter such as plant material and microbes. The quantity of organic material (which is important in subsequent treatment), and whether it is alive or dead, will depend on seasonal and climatic conditions, including temperature and rainfall. Categories (b) and (c) usually offer water that contains noticeable quantities of hardness salts that have dissolved in the water as it permeates through the ground. The concentrations of hardness salts in a water supply, which obviously affect water quality, are also used to categorize waters (see later). The quantities of suspended solids and dissolved oxygen in deep well waters (c) are normally consistently very low, whereas in surface waters, (a) and (b), these quantities tend to be higher, depending on climatic conditions such as temperature, rainfall and season (falling leaves).

3.1.3 Seawater

The oceans are by far the largest source of water in the world, accounting for about 97% of the water volume on our planet. Seawater is not just a solution of sodium chloride: it also contains many other ionic species (notably those listed in Table 2.2), together with other materials, both dissolved and held in suspension. The *salinity* of seawater — a measure of the total concentration of dissolved salts — varies from place to place, even in the surface waters of the open oceans. But close to land, local conditions can have a more marked effect. For example, estuarine water has a reduced salt concentration simply by virtue of dilution by the river water. Estuarine water in tropical regions contains high concentrations of tannins and decaying vegetation from rainforests, while in uncontrolled industrial areas there may be severe pollution and contamination from commercial activities. Along the coast, even many kilometres from a large river such as the Amazon, the sea is often turbid because of vast amounts of mud swept down by the river.

At coastal power and industrial plants, seawater is pumped through culverts and pipes for use as a coolant to condense steam and for other cooling purposes. The water is then returned to the ocean at a higher temperature and quickly dispersed by dilution. However, the use of seawater as a coolant can present problems. Some seawater holds sand in suspension, especially with high swells, and may contain vegetation and animal life, which can interfere with an industry's cooling water circulation system. Sand can have an abrasive action on pipes and pumps.

Living matter can grow and block pipework and so it has either to be prevented from entering the system or it has to be killed chemically on entry to a site. There are restrictions on the amounts of chemicals employed in this way, as the residues could be discharged to the environment with the returned warmed water, as discussed in Section 5. Near an estuary, the high mud burden of seawater can result in mud settling inside an industrial cooling system, thereby reducing heat-transfer efficiency, whilst also offering an ideal habitat for marine life. For example, certain anaerobic bacteria in mud thrive by removing oxygen from sulfates, leaving sulfides that can cause corrosion of the copper pipes often used in cooling water installations. This problem may be serious enough to warrant the installation of equipment to remove the mud, by either physical or chemical means.

Apart from its role as a coolant, seawater is not very attractive for many potential users because of its salt content. Nevertheless, in regions of the world that have minimal supplies of freshwater but have plentiful supplies of seawater without too much mud in suspension, purification by desalination may be practicable, especially if a relatively cheap source of energy is also available.

3.2 Water categorized by hardness and softness

The quality of waters can also be categorized by the amounts of dissolved so-called 'hardness salts' they contain. Certain solutes are described as causing hardness and water is termed hard because, when such water is boiled vigorously, a physically hard scale is deposited on the heat-transfer surfaces. There are two ways in which this scale deposition can occur.

The first mechanism is simple evaporation in which the water is merely driven off by the boiling process, leaving behind all solid solutes as dried-out deposits. Incidentally, hard scale has a much lower thermal conductivity than water, thereby reducing the efficiency of the heating process. This is why it takes longer, and so is more expensive, to boil water in a 'furred-up' kettle than in one without scale.

The other mechanism by which hard scale is deposited is a little more complicated as it involves chemical decomposition. Rain with a low pH will dissolve some calcium carbonate, $CaCO_3$, in chalky and limestone areas, producing calcium, Ca^{2+} and hydrogen carbonate, HCO_3 ions. A similar outcome results when rainwater comes into contact with magnesium carbonate, $MgCO_3$. Limestone is predominately calcium carbonate, whereas dolomite comprises calcium and magnesium carbonates. The effect of rainwater containing dissolved carbon dioxide on these two carbonates is given in Equations 3.1 and 3.2:

$$CaCO_3 + H_2O + CO_2 = Ca^{2+} + 2HCO_3^- \quad (3.1)$$

$$MgCO_3 + H_2O + CO_2 = Mg^{2+} + 2HCO_3^- \quad (3.2)$$

The result is a solution of the hydrogen carbonates of calcium and magnesium and this confers **temporary hardness** to the water. If water containing these solutes is boiled, even gently, these hydrogen carbonates easily break down into water, carbon dioxide and the corresponding carbonates (Equation 3.3). The carbonates that are produced precipitate on heat-transfer surfaces such as electric kettle elements. In other words, the reverse of the previous reaction takes place on heating water with temporary hardness.

$$Ca(HCO_3)_2 = H_2O + CO_2 + CaCO_3 \quad (3.3)$$

Thus, temporary hardness can be removed from water by boiling, even though the result is the production of troublesome calcium and magnesium carbonates. However, the carbonates (and sulfates) of calcium and magnesium cannot be destroyed by boiling, and so they are termed **permanent hardness** salts and the water is regarded as being permanently hard. Permanent hardness is a rather unfortunate term because, with suitable chemical treatment, it may also be removed.

Question 3.1

Is there likely to be a difference between the hardness of water flowing over limestone-rich soils where that water has come from rain falling directly on the soil and water that has originated via throughfall in forested areas?

Soap removes dirt from soiled fabrics, particularly clothes, because most dirt is held on cloth by films of lipids (oils, fats and greases) and soap can form an emulsion with these in water. Lifting a lipid from a cloth as a soapy emulsion suspended in the water also enables the dirt to float off the cloth. However, when you use soap in a hard water area, the soap initially reacts with the hardness salts to form insoluble precipitates (a scum); only after all the hardness salts have reacted, or have effectively been used up, is it possible to produce a lather. Indeed, until the mid-20th century, water hardness was actually measured by the amount of soap required to produce a lather under specified conditions. In commercial laundries, and to a lesser extent in domestic washing machines, scum in hard water areas can cause a problem when it is entrained in fabric fibres and becomes difficult to rinse out.

The chemical reaction involved in scum formation is:

$$\underset{\text{stearate ion in solution (soap)}}{2C_{17}H_{35}CO_2^-(aq)} + Ca^{2+}(aq) = \underset{\text{insoluble calcium stearate (scum)}}{(C_{17}H_{35}CO_2)_2Ca(s)}$$

The product of the reaction, calcium stearate, is insoluble in water. Detergents are similar in action to soaps but the sulfonate and other groups in the molecules do not form insoluble compounds with calcium and magnesium cations in water, hence no scum. When washing soiled clothes, the ionic end of the detergent molecule (the hydrophilic, water-loving, end) is attracted to the water, while the hydrocarbon end (the lipophilic, lipid-loving, end) goes for any grease that is on the fibres. The result is that the hydrocarbon end of the detergent molecule embeds itself in the grease and grime while the ionic end, firmly in the washing water, and particularly if the water is agitated, tugs on the whole molecule thus pulling the grease away from the fibres. The detergent molecules (with grease and dirt attached) remain in suspension in the water, the grease does not return to the fabric and there is also no scum.

When first introduced, detergents were manufactured with many cross-linkages in their long-chain structure, which impeded biodegradation; this resulted in massive volumes of foam forming at sewage treatment works in the 1950s and 1960s. Subsequently, the biodegradability of the hydrocarbon part of the detergent has been improved, and the replacement of sulfonate groups by sulfate has helped further. Many of the detergents used today are alkyl sulfates.

3.3 Summary of Section 3

1. Permanent hardness is caused primarily by calcium and magnesium carbonates. Temporary hardness arises through the hydrogen carbonate salts of these metal cations.
2. Water draining from peaty, moorland areas is generally soft but surface waters draining from carbonate-rich areas (surface or underground) are hard.
3. Soaps produce an insoluble scum with hard water but detergents do not.

4 Abstraction of water from the environment

For centuries, the water needs of communities have been met from a variety of sources, in the first instance from the most accessible or abundant supply. Originally settlements were set up adjacent to a supply of freshwater so that many towns and villages and religious communities grew up on the banks of rivers or on the sides of freshwater lakes. After the water had been used, much of it was discharged as effluent. Consequently, over the years, water supply and sewage disposal became related activities, though unfortunately not always related responsibilities.

If a stream could not satisfy the immediate demands of the community, it might be dammed in order to build up reserves; hence artificial lakes and reservoirs were constructed. As communities moved away from rivers and lakes seeking alternative lands for grazing their flocks, or perhaps settling where trading routes crossed, their water needs were met by carrying, or devising a method of conveying, the water from the river or lake to the settlement. If geologically feasible, an alternative strategy was to dig wells or sink boreholes into the underlying water-bearing rock strata, although this obviously required a greater effort. In this section we shall look at some of the methods that may be used to procure supplies of water for human needs.

4.1 Abstraction from rivers

In Roman times, water was delivered via aqueducts (Figure 4.1) to towns and even directly to a few homes, normally to a fountain or pool in a courtyard.

However, for the vast majority of the population water was collected from the local river. Those who had the means paid others to fetch and carry it for them,

Figure 4.1 A Roman aqueduct.

and the specific trade of water carrier developed in many large towns. In 15th century London, apprentices had 'to carry tankards to serve their masters' houses with water fetched from the River Thames or the common conduits'. Until this period, the quality of a river water, was usually suitable for human consumption. Over the following 300 years in London, several private water companies were formed to provide water, almost entirely from river sources around the capital. The first water company, set up in 1582, took the water directly from the Thames in the centre of London but a plague outbreak in 1603 made the inhabitants suspicious of the quality of the water provided. Further water companies were established in the 1600s, three in the 1700s, the others during the 1800s.

An ambitious plan by the New River Company, formed in 1603 and financed by Hugh Myddelton, involved constructing an aqueduct 40 miles long to convey water from Hertfordshire springs to central London. An open conduit took the water into London and then wooden pipes were used to take this freshwater through the streets and finally into private houses in lead pipes. The City of London's Corporation originally refused to commit any money to this wild scheme. However, in 1613 the King (James I) came to the rescue and in that year Hugh's brother, Sir Thomas Myddelton, became Lord Mayor and the first supplies of New River water splashed into the Round Pond at Clerkenwell. Much of the original aqueduct remains today, almost 400 years later, though now shortened to 27 miles, and still provides a considerable volume of water to Londoners. Other water companies supplying water to London used the River Thames as their source, distributing the water directly to their customers.

By the beginning of the 19th century, London water quality — apart from that of the New River supply — was so unsatisfactory that a Royal Commission was set up; this recommended changing supply sources. Cholera epidemics in 1848 and 1849 were attributed to impure water and in 1850 the General Board of Health recommended that the River Thames should be abandoned as a source of supply. In spite of the death of Queen Victoria's husband, Prince Albert, of typhoid in 1861, it took until 1902 for the politicians to pass the Metropolis Water Act by which all the existing water supply systems came under the control of a Metropolitan Water Board. At the end of the 1800s, eight companies were still supplying London's needs, mostly from the River Thames and the River Lea. These companies were all incorporated into the Metropolitan Water Board in 1904.

Sir Alexander Houston, Director of Water Examination to the new Metropolitan Water Board, discovered that when water was stored there was generally a reduction in the bacterial content. Storage alone could, therefore, be used to improve water quality in terms of potability; a novel method of internal water treatment had been conceived. This important observation encouraged the Metropolitan Water Board to construct large reservoirs, not only increasing the combined storage capacity for London from 18 000 million litres in 1904 to 412 000 million litres in 1950, but also providing a method of improving water quality without adding chemicals or even using mechanical methods such as filtration. Under exposure to sunlight and air and with the low flow rates from one end of a reservoir to the other, enabling clarification by precipitation and sedimentation to take place, many pathogenic microbes and impurities were eliminated. Today, the water is also filtered and chemically treated before being conveyed to consumers.

4.2 Abstraction from natural lakes

In contrast to London, Glasgow depended on some 30 public wells for its water supply until 1800. These produced relatively hard water that also was becoming increasingly susceptible to contamination as the city population began to accumulate in high-density housing. However, in 1800, spring water was piped into a tank in the city centre, and the uncontaminated water was sold at a halfpenny a stoup (almost 30 litres).

A scheme for conveying even softer and equally uncontaminated water to Glasgow from Loch Katrine, a large natural lake in the Perthshire Highlands nearly 50 kilometres away, was debated for almost 60 years. Finally the scheme was approved, work was instigated and in October 1859, the completed system was inaugurated by Queen Victoria. The quality of the water was so high for drinking purposes that no treatment was necessary, even 100 years later, and the water was so soft that there were major savings in the quantity of soap used by Glaswegians. In fact, Loch Katrine was naturally utilizing the storage method of internal water treatment that Alexander Houston 'discovered' some 50 years later in London, and which led to the construction of artificial reservoirs by the Metropolitan Water Board. The water levels in Loch Katrine were raised in 1880, and then again in 1913 to a depth of just under six metres to cater for the increasing population of Glasgow.

4.3 Abstraction from artificial lakes and reservoirs

In common with many large conurbations, Edinburgh relies on artificial lakes and reservoirs for its water supply, as there is no suitable river or natural lake in the vicinity. An ideal site for an artificial reservoir (Figure 4.2) is a valley that, after flooding, is able to hold about six months' supply for its consumers. A dam, of either earthwork or masonry, is needed together with provision for water to be passed downstream when the reservoir is full. When calculating the size of a potential reservoir, the following factors are considered: the average local rainfall, the size of the drainage or catchment area, the available rainfall (the amount of rain that the drainage area can actually make use of), the riparian rights of downstream users in terms of the volume of compensation water (traditionally a third of the amount that would have passed downstream if the reservoir was not built) and fishing rights, taking into account that some fish swim upstream to spawn and a reservoir may impede their travel. Two reservoirs, one 56 km from the city and another 24 km away, were constructed for Edinburgh. While not all that far apart, the annual available rainfall for each is quite different, one being 1016 mm, the other 1651 mm; and so the capacity of the latter reservoir could be less than of the former. In the event, the capacity of the reservoir in the former area was 14 000 million litres and in the area with the higher rainfall was 9 000 million litres.

Figure 4.2 Haweswater Reservoir in the English Lake District, which together with Thirlmere Reservoir provides water for Manchester.

4.4 Conveyance of water

Today, in the developed world, we simply turn a tap, and water of a quality suitable for drinking just flows out. Apart from a few large cities in Britain, it was not until the 1850s that enterprising individuals began to form water companies to convey water to domestic dwellings. In the mid-19th century, most British people had to draw and carry their water from a public source, often a well that the local authority was obliged to provide. Only those who could afford to pay were able to have their own supplies, though often delivered to only one tap in the house. Even in the mid-20th century over half of the houses in Great Britain had neither baths nor hot water systems and showers were unheard of. Many thousands of homes at this time were using earth closets in the back garden or at the bottom of the yard.

In England and Wales, the Waterworks Clauses Act of 1847 gave statutory powers (with obligations) to municipalities and companies to acquire water rights and provide water supplies. They were required by the legislation to lay pipes in streets and convey water under sufficient pressure to rise at all times to the highest homes in the area. Borough Councils were permitted to fit fire plugs to these supply lines for fire-fighting purposes. The profits of the water companies were limited to a certain percentage. For years, many councils chose to purchase water from private companies rather than using ratepayers' money for risky new schemes; but by the 20th century most water utilities were municipal undertakings. Towards the end of that century, political policies and financial interests resulted in all the water utilities reverting to being owned by private companies and individual shareholders.

Originally water was transferred from source to consumer, not quite on a one-to-one basis but there was little need to consider a system by which water was conveyed from source to an intermediate or service cistern or reservoir and then pumped or gravity-fed through a distribution system to the user. Gradually, however, it was realized that having holding reservoirs offered not only better management of water provision but also an opportunity to purify the water locally. Progress in Britain has been rapid over the past 50 years, but there are very many places in the world today where water supplies and quality are similar to those of 19th century Britain.

4.4.1 Conveyance from source in aqueducts

Water is conveyed along aqueducts from the large storage reservoirs to a service reservoir near to a community or industrial complex. The aqueducts are normally a combination of tunnels or pipes fabricated in a variety of materials. Centuries ago, when an aqueduct crossed a valley, huge arched brick or stone viaducts were built to convey the water, but today pipes that follow the valley's contour are used; this arrangement is termed a siphon. The water velocity in tunnels and across masonry viaducts has to be low (about one metre per second) to minimize erosion; while in pipes, where there is no great erosion problem, the water velocity depends on the hydraulic gradient, which is the difference in level between the inlet and the outlet. The greater the gradient, the greater the volume of water that can be conveyed through pipes of a similar diameter without pumping.

Along a water pipeline, provision is made at suitable intervals for the pipe to be completely drained if necessary. When refilling a pipeline, an appropriate number of air vents and valves are needed along its length to remove airlocks. The contractor who built the first piped water supply in Edinburgh in 1676 did not understand this requirement. At the grand opening ceremony, water failed to emerge from the three-inch diameter lead pipe laid along three miles from the Comiston Springs. In confusion, the contractor fled on horseback but a knowledgeable bystander simply pierced the pipe at a few high points and the water gushed out.

In any aqueduct the quantity of water flowing is the same throughout its length. In a small diameter pipe the water flows more quickly; conversely, if the water velocity has to be slow, say to minimize erosion, the pipe or tunnel diameter has to be larger. Cast-iron pipes were manufactured from the 18th century, connected by flanges using sheet-lead washers. Spigot and socket joints were next introduced with rope yarn forced into the join and molten lead poured in and driven hard with caulking tools when cool. Steel pipes followed, some lined and coated with bitumen to minimize corrosion, although the lining had to be continuous to be effective; accordingly, this technique was applied only to pipes with a diameter large enough for a person to crawl inside to use flame burners to melt and work the bitumen. Reinforced concrete pipes have been used for large diameter pipes but, in general, in shorter lengths than for steel pipes, requiring more joints. Asbestos cement pipes were used for many years, being cheaper and lighter than concrete and not subject to metallic corrosion, before concerns were expressed over the use of asbestos in certain situations.

In recent years, water has been conveyed over very long distances from source to service storage and then to the customer; this means that nowadays the water from a local tap may have few characteristics of the local geology (Table 4.1).

Table 4.1 Lengths of aqueducts from water sources to the areas they supplied in the UK.

Area supplied	Source	Aqueduct length/km
Manchester	Lake District	170
Birmingham	Welsh mountains	117
Liverpool	Welsh mountains	109
Birkenhead	Welsh mountains	75
Newcastle	Cheviot Hills	61
Glasgow	Loch Katrine	56
Edinburgh	Southern Uplands	54
Cardiff	Welsh mountains	48

4.4.2 Service storage and distribution

Service storage facilities are usually covered tanks or cisterns situated at elevations sufficient to allow the water to fall under gravity to the required area. The size of the service storage has to be large enough to hold the difference between the average daily demand and the maximum demand. In some places, these tanks may need to be artificially elevated in water towers. The water may require treatment or further treatment before being distributed to the users, depending on the source or sources of the water, any treatment it has received between the source and the service storage, the method by which and distance over which it has been conveyed from the source to the service storage, and the material from which the aqueduct is made.

The pressure of the water has to be sufficient to give an adequate supply to each user, but not so great that fittings are damaged. Most users are provided with gravity-fed water from local storage tanks, either serving a community from a water tower or serving individual buildings from storage tanks in their own roofs.

Water in the UK was traditionally supplied to domestic customers in return for a fixed payment on a periodic basis, once a month or every six months, irrespective of the amount of water consumed. For many years, industry has purchased the water required by volume and supplies to each industrial user have been metered by the water companies and authorities. Currently, all customers in the UK are being encouraged to take more account of their real water needs and their actual water consumption. Many domestic users now have meters fitted to their supply pipes and, as an incentive, some water companies are offering light-user schemes with reduced prices for low water consumption. The quality of the water supplied remains of drinking water standard.

4.5 Summary of Section 4

1. Water for human consumption was originally abstracted from rivers and used without further purification.
2. The development of conurbations often resulted in the contamination of local (surface or underground) water sources and water had to be piped in sometimes over considerable distances.
3. There have been developments in materials used to structure pipes, major criteria being that the pipes should be resistant to corrosion and the pipe material must not contaminate the water.
4. For a given supply of water, pipe diameter has to be sufficient to give a speed of flow that does not result in significant erosion of the pipe material. As far as is possible, pipe routes are designed to maximize gravity flow and to minimize pumping.

5 Discharge and return of 'used' water to the environment

Huge volumes of water are discharged or returned to the environment by domestic and industrial consumers after they have used it for their own particular purposes. In the UK, over 11 000 million litres of water per day are put into sewers alone. In the vast majority of situations, the quality of this effluent differs substantially from that of the water initially provided to the user. In some cases, the effluent contains by-products for which the consumer does not have a use, and so is not concerned about. The effluents produced by all domestic users, and by some commercial users, fall into this category. By-products of particular concern are the result of changed concentrations of organic chemicals such as hydrocarbons, phenolated compounds and oils, or inorganic chemicals such as heavy metals, mineral acids or chlorine. Effluents can also be toxic or radioactive and may lead to unacceptable cloudiness, odour or taste. However, many industries are now aware of both their responsibilities and their legal obligations and make suitable provision to remove the by-products from their waste water.

It is certain that the more sophisticated methods of analysis currently available have enabled us to detect very small amounts of chemicals that in the past we did not even realize were present in many waters. Now that we know these substances are there, even in minute concentrations, we have begun to worry about them. The quality of water may not have changed over the years but our ability to discover what is in it has.

5.1 Pollution and contamination

If we could be sure that all domestic and industrial water users removed all of their by-products from the water they have used before returning it to the environment, there would be no need to monitor and control effluents and no concerns for pollution or contamination. However, this is not practicable, especially for domestic water users; even industrial users who have installed sophisticated treatment units sometimes encounter problems through faulty design, improper operation or inadequate maintenance of their equipment (Figure 5.1).

A useful distinction between **pollution** and **contamination** is that pollution is the introduction of damaging (to living organisms) loads or concentrations of material or compounds while contamination is the introduction of new material and compounds (not necessarily at damaging levels) to the system.

National and international advice and constraints are published on what a user can or cannot, or should not, leave in waste water. The first legislation in the UK dealing with pollution of rivers by sewage and industrial wastes was introduced in 1876. Legal restrictions are imposed today in the UK on discharges to the environment and on the pollution of controlled water, through regulations such as the EC Urban Waste Water Treatment Directive 1993 and Section 85 of the Water Resources Act 1991, the Water Industry Act 1991 and the Land Drainage Act 1991. If water is returned to a river, the main concern is for the downstream uses — irrigation, cooling, fishing, swimming, drinking, etc. Advice and constraints have varied over the years as our knowledge and understanding of health and hygiene have developed, as our means of sampling have altered and as our analytical techniques in detecting chemicals have improved.

One convenient check on the effect of discharged materials on water quality in the environment is to measure the biological oxygen demand (BOD) in the water. BOD is a common measure of organic pollution, representing the amount of biologically degradable substances in the water or effluent sample. A test sample is stored for five days at 20 °C and the amount of oxygen taken up by the microbes present is measured in grams per cubic metre (milligrams per litre). A clean mountain stream has a BOD of about 0.5 mg l^{-1} but water containing untreated sewage can have a BOD as high as 25 or even 250 mg l^{-1}. The HERMES model can be used to follow the progressive change in BOD (a non-conservative pollutant) along a river.

Figure 5.1 The manufacturing processes for paper, paints, detergents, drinks and vehicles all use large quantities of water, which have to be returned to the environment.

5.1.1 Sources of pollution and contamination

Urban waste water usually contains not only domestic effluent from sinks and baths with sewage from toilets but also rainwater from roofs and roads. In some countries, two pipework systems are installed, one for domestic waste water which is directed to a treatment plant, the other for rainwater which requires little or no treatment before being discharged to a river. Industrial waste water is often treated within the premises. In the UK, over 95% of the polluting load of sewage is successfully removed by municipal treatment plants before the water is discharged to inland waters, even though domestic sewage accounts for over 25% of all pollution.

Agrifood industries include piggeries, slaughterhouses for livestock and poultry, tanneries, glue and gelatin factories, starch and food producers, and the waste water from these industries is inevitably contaminated. Gas scrubbing, the bubbling of gaseous effluents through water, is used to remove hydrogen fluoride and hydrogen cyanide from blast furnace gases in the steel industry, to remove hydrogen chloride produced in the incineration of household waste and to remove sulfur oxides from smoke, following the combustion of heavy oils and coal. The resulting waste waters are significantly acidic. The coal and metal mining and wood industries leave vast quantities of silt and sawdust in their effluent water. Neither silt nor sawdust is directly poisonous, but silt can cover a lake bottom used as feeding grounds by aquatic organisms and the bacterial decomposition of sawdust removes dissolved oxygen from the water.

Both pollutants and contaminants can come from point sources (a distinct location) or diffuse sources (an area). Whilst these effluents are mostly the results of (i) increasing urban development and its sewage, (ii) different industries, and (iii) changing farming practices, it is important to recognize that much of the material discharged into UK rivers and lakes from over 12 000 point sources is under approved licences. Table 5.1 identifies some sources of typical water pollution into UK rivers and the major polluting agents.

Table 5.1 Typical sources of water pollution and parameters affected.

Source	Pollutants
domestic sewage	ammonia, nitrate, phosphate, suspended solids, BOD
agricultural run-off	nitrate, phosphate, pesticide residues
chemical industry	ammonia, phenols, non-biodegradable organics, heat, BOD
iron and steel manufacture	ammonia, cyanide, phenols, sulfides, pH
coal mining	iron, suspended solids, dissolved solids, pH
metal finishing	copper, cadmium, nickel, cyanide, pH
dairy products	BOD, pH
oil refining	ammonia, sulfides, phenols, oil, heat
power generation	heat

5.2 Oxygen levels in water

At around 15 °C, a minimum of about 0.03 litre of oxygen per litre of water is required for the survival of aquatic animals. At lower oxygen levels, life in rivers and other aqueous habitats will alter in some way or other. The concentration of oxygen dissolved in water falls as the temperature of the water rises. If significant volumes of hot water are discharged into slow rivers or lakes by industrial users, oxygen levels may decrease so much that fish can no longer survive. Other factors can also reduce oxygen levels in water.

Ammonium compounds from sewage, or discharged to a river from factories and farms, are oxidized by bacteria in water to nitrates and nitrites, using dissolved oxygen. The nitrites are further oxidized to nitrates, thereby removing more oxygen from the water. Nitrates are nutrients for algal growth near the water surface, particularly when the water flow rate is low. The region of water into which light can penetrate, enabling photosynthesis to take place, is termed the photic zone (Block 4, Part 2 *Cycles*, Section 4.2) (Figure 5.2).

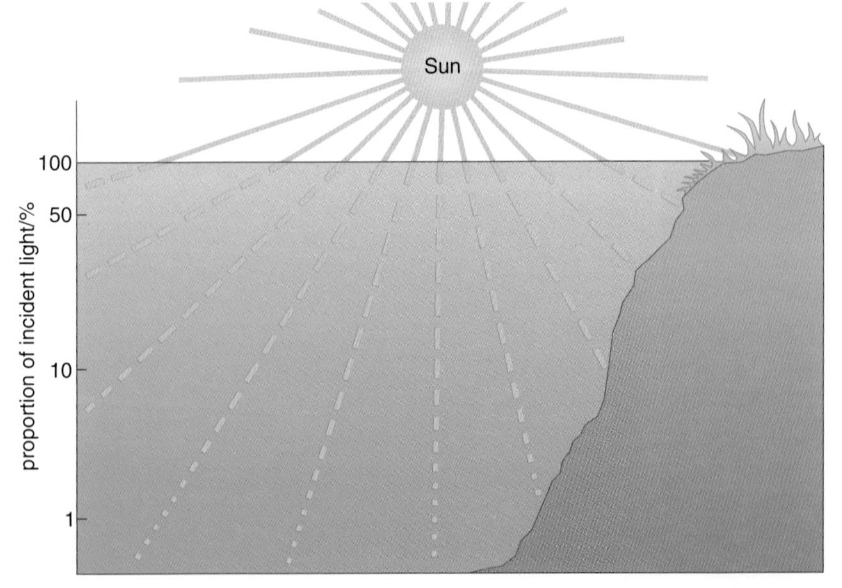

Figure 5.2 Penetration of light through clear water.

Algal growth reduces the amount of sunlight that can penetrate into the water, so reducing the depth of the euphotic zone (Figure 5.3). The reduced level of photosynthesis would lower oxygen levels in the water. When the algae die, their decay consumes more dissolved oxygen and the water may no longer be able to support respiring life. These detrimental effects can be reduced if water is flowing, thereby increasing the oxygenation rate from the atmosphere.

Figure 5.3 Algal growth on the surface of a water channel.

5.3 Discharged materials

Untreated sewage discharged into seawater can be a problem if it gathers around bays and inlets. However, when it is carried by tides and coastal currents to deeper water, sewage is quickly decomposed and becomes relatively harmless.

The municipal treatment of sewage normally removes the bulk of its organic matter. However, this process liberates phosphates from the sewage to add to the phosphate detergents already present in the waste water. Phosphates may then join nitrates in rivers and lakes to promote plant and algae growth and possible eutrophication. Lake Annecy on the Swiss/French border was rescued from this situation by diverting the municipal sewage outfall that originally discharged directly to the lake. To prevent this situation recurring in Switzerland, a limit of $2\,mg\,l^{-1}$ of total phosphates is imposed on all discharges of treated waste water into its lakes.

Industrial waste may contain cyanides and various heavy-metal compounds of lead, mercury, copper and zinc. These chemicals are potentially dangerous, even in low concentrations, because they can accumulate in fish and other aquatic organisms and then spread along food chains to other animals, including humans. The problem is more acute in rivers, but even at sea these wastes can accumulate in shallow-water fisheries and so enter the human food chain.

Agricultural wastes include manure that was traditionally spread on land. Modern methods involve raising poultry, cattle and pigs in buildings, resulting in the accumulation of manure, which may produce runoff into ditches and thereby pass into streams and rivers. As the manure decomposes, it reduces oxygen levels in the water creating problems we have already mentioned. Insecticides and fungicides in crop sprays and fertilizers containing nitrates and phosphates may be washed into local ponds and rivers, creating chemical imbalances and also disrupting food chains.

Oil spillage into oceans and river estuaries from tankers and offshore oil rigs may poison seabirds directly or damage their feathers affecting flight and 'waterproofing'. If oil is washed ashore, all forms of life on the rocks and beaches will be harmed. Well-intended cleaning of birds' plumage with detergents may release phosphates into the sea, causing more damage than if the oil had been allowed to be broken down by normal microbial action. Even at 1 ppm, detergents kill most shore life. On the other hand, if floating oil is covered with chalk, it sinks to the sea-bed and is destroyed there by oil-digesting anaerobic bacteria.

5.4 Summary of Section 5

1 Biological oxygen demand can be used to estimate the concentration of organic matter in water.

2 Domestic sewage treatment plants are not generally equipped to remove some of the contaminants of used water. This particularly applies to some industrial contaminants such as mercury, where very small concentrations can be damaging to life. In these cases, such contaminants must be removed as part of the industrial processing on site.

Water treatment

6

6.1 Goals of water treatment

For water to provide a suitable habitat for organisms, it must contain dissolved oxygen; for water to be used in circulating systems constructed from metals, for the transfer of heat or pressure, all the dissolved oxygen has to be eliminated to prevent corrosion. Land-based plants and animals can make use of oxygen from the atmosphere. Many organisms have migrated to areas where the water quality is more acceptable for their particular needs; some have adapted to take advantage of the available water. Humanity has taken a different approach to the problem in the past 200 years, and has attempted to adjust the quality of a raw water to meet varying needs, through some form of treatment. The goal of any water treatment is to serve the needs of the majority of the users although, at times, there are bound to be conflicts between the requirements of different customers drawing from the same treated water supply.

The supplied water (treated if necessary) has to be potable and be free of colour so that it is suitable for washing the body and clothes. Domestic, hotel and other users with water-filled central heating systems may prefer to have few hardness salts in the water to minimize maintenance and running costs (so some artificial softening may be needed). Electronics companies may also require hardness-free or demineralized water in their washing plants. Industries using steam, particularly at high pressures, are also likely to use demineralized water in their boilers to optimize heat transfer and eliminate scale formation. Such industries may also have to remove oxygen and adjust the pH to minimize corrosion. By contrast, water for cooling in industry, irrigation of a golf course or rinsing copper mine ores does not necessarily demand the same quality of water as for manufacturing pharmaceutical drugs.

In this section we shall look at some users' requirements and methods of water treatment. Apart from cooling-water systems, you will see that in most instances public water supplies are treated to drinking water standard and users who require water with less solutes install facilities on their premises to further treat the water to meet their particular needs.

It is worth noting that a therapeutic spa is one place where water treatment is rarely applied as the attraction is to offer to the visitors the natural water with all its (beneficial) 'impurities'.

6.1.1 Drinking water standards

The regulatory framework

The potentially damaging substances that are likely to appear in public drinking water are quoted in European and UK legislation, which set limits on permissible concentrations. These limits range from zero, for potentially dangerous microbiological organisms such as faecal bacteria, to various concentrations for other parameters of health and aesthetic significance. A *Directive on the Quality of Water Intended for Human Consumption* was adopted by the European

Commission (EC) in 1995 and came into force in 1998. The European Member States, including the UK, were required to implement legislation by December 2000, and to comply with the majority of the standards in the Directive within three years. The 1995 Directive was a revision of a 1988 Directive and now sets criteria for drinking water standards that are generally in line with the 1993 World Health Organization (WHO) guidelines.

Drinking water may be supplied from a distribution network, tanker, bottle or container. The WHO guideline values represent the concentration of a constituent that does not result in any significant risk to the health of a consumer, usually over a lifetime of consumption. Values are set with substantial safety margins so that a limited or temporary departure from them does not necessarily give rise to any risk to health.

The current regulatory regime in the UK was introduced in 1989 when the Water Act 1989, later consolidated into the Water Industry Act 1991, established a framework of new bodies responsible for water and sewerage services in England and Wales to replace the former Water Authorities. At the same time, mechanisms were identified to ensure the proper provision of water services.

For England and Wales, the limits for potentially damaging substances are monitored by water companies in some three million tests a year. Methods of monitoring are not specified in the EC Directive, only that they must meet certain performance standards. This enables methods to be modified in line with technical and scientific progress, without changing the Directive. Water companies are required to take a specified minimum standard number of samples for each parameter within their supply zone, according to the population served by the zone. This 'self-monitoring' role is subject to checks by local authorities and the tests are audited by the Drinking Water Inspectorate.

British water companies are under a duty, when supplying water to premises for domestic (drinking, washing and cooking) or food production purposes, to provide only water that is wholesome at the 'time of supply'. The quality of water as it is actually used for food production is additionally the subject of different controls under the Food Safety (General Food Hygiene) Regulations. Time of supply is defined as when water passes from a company's pipe into a consumer's pipe. Particular mention is made of amounts of bromate, trihalomethanes and lead in the water. Water of low pH value may dissolve lead and the maximum level of lead in drinking water is set at $10\,\mu g\,l^{-1}$ in the current Directive (compared with $50\,\mu g\,l^{-1}$ in the 1988 Directive) to protect fetuses and infants from the neurotoxic effects of lead that can contribute to learning and behavioural problems.

Under the 1991 Water Act, 'wholesomeness' is described by reference to standards (maximum admissible concentrations and minimum required concentrations) set out in the EC Drinking Water Directive and also to 11 national standards. In total, numerical standards are set for 55 parameters and descriptive standards for a further two. Ten parameters are mentioned below. The UK Regulations, in some instances, impose higher standards than the current EC Directive. This is generally because the Regulations are based on values in the 1988 Directive, some of which have now been relaxed by the EC (while others have been tightened) but, to meet local requirements, have been retained within the UK. For example, water that has been softened before supply should retain a hardness of at least 60 mg of calcium per litre (or its magnesium equivalent) at the consumer's tap — although water that is naturally soft does not have to be hardened to this level. This soft/hard water attitude is based on the lead dissolution observations mentioned above, and also the UK 1994 and subsequent research on cardiovascular disease, which shows consistent evidence of low incidence of heart disease-related mortality in areas of naturally hard water. Sodium levels are also omitted from the 1998 Directive, although the UK Regulations set a limit of 150 mg l^{-1} Na. Tetrachloromethane levels are not in the 1998 Directive, but a limit of 3 $\mu g\, l^{-1}$ is included in the UK Regulations as some groundwater sources for drinking water have been found to carry this.

Table 6.1 Maximum concentration levels for some substances in the current UK Regulations (Statutory Instrument No. 3001, 1991).

Substance	Maximum concentration/mg l^{-1}
nitrate (NO_3^-)	50
fluoride (F^-)	1.5
copper (Cu)	0.05
arsenic (As)	0.05
cadmium (Cd)	0.005
lead (Pb)	0.05
selenium (Se)	0.01
mercury (Hg)	0.001
polycyclic aromatic hydrocarbons	0.0002
total pesticides	0.001

There is an overall requirement in both the Directive and Regulations that public drinking water must not contain any element, organism or substance, whether alone or in combination, at a concentration or value that would be detrimental to public health. Drinking water is regarded as wholesome, provided it meets these standards. All standards apply to water at the time of supply (see above); there are national standards that apply to water leaving treatment works and to water held in service reservoirs within the distribution system.

The Regulations require water companies to maintain records on the quality of water supplied in their zones. Registers of results must be open for inspection and copies of information on the register must be provided to any person on demand. As an individual you are entitled to a copy of information, free of charge, regarding your own water supply. Annual reports must be sent to local authorities and water companies must also publish an annual report containing information on the quality of the water supplied.

The Drinking Water Inspectorate is responsible for initiating enforcement action. Under Section 18 of the 1991 Act, enforcement action must be taken when there is evidence that an appointed company is contravening, or has contravened and is likely to contravene again, any statutory requirement enforceable under that Section (Table 6.1). However, enforcement action is not required if the contravention is trivial, or that the company has already given an undertaking to take appropriate steps to secure or facilitate compliance with the requirements.

6.1.2 Swimming pool water

Unlike drinking water, for which there are international guidelines on quality and, in the UK at least, also legal requirements in the form of Regulations, for swimming pool water there are only recommendations, although other aspects of a public swimming pool are subject to legislation. However, there is a European Directive on the quality of seawater in which people are likely to swim, the Bathing Waters Directive. Details of the recommended quality of swimming pool water were first published in 1929. Today, however, swimming pool water quality and treatment are based on the research and recommendations of an independent Pool Water Treatment Advisory Group, established when government withdrew from the field (Table 6.2). Current policy is to educate pool owners and managers rather than prescribe specific approaches.

Table 6.2 Comparisons between drinking water and swimming pool water. (max = maximum; min = minimum)

Parameter	Water standard supply values (maxima except where indicated)	Swimming pool recommendations (maxima except where indicated)
pH	5.5 (min) 9.5 (max)	7.8 (min) 9.2 (max)
conductivity	1500 μS cm^{-1}	4250 μS
turbidity	4 FTU*	0.5 FTU*
temperature	25 °C	27–30 °C
sulfate	250 mg l^{-1}	300 mg l^{-1}
sodium	150 mg l^{-1}	none
chloride	400 mg l^{-1}	none
magnesium	50 mg l^{-1}	none
calcium	250 mg l^{-1}	18–30 mg l^{-1}
total hardness	60 mg l^{-1} Ca (min) (if softened)	30 mg l^{-1} Ca
trihalomethanes	100 μg l^{-1}	100 μg l^{-1}

* FTU = formazin turbidity units (a measurement unit for turbidity).

Legal adherence to swimming pool standards in the UK is covered by several Acts of Parliament from 1974 to 1998. The responsibility for enforcing the legal requirements, under the Health and Safety (Enforcing Authorities) Regulations 1998, lies with the Local Authority Environmental Health Officers for public pools in their areas except those pools owned and operated by that Local Authority. All other public, educational establishment and Ministry of Defence pools are the responsibility of the Health and Safety Executive.
Public swimming pools that were formerly river diversions or natural lakes are no longer provided or sanctioned by Local Authorities in the UK. Raw water that is used to initially fill a public swimming pool complies with drinking water standards (Table 6.2). However, some potable waters with low pH and/or low hardness appear to attack pool tile grout and circulation systems; furthermore, swimming pool water, even if initially wholesome, is constantly being polluted by bathers.

Accordingly, pH is controlled and potentially harmful bacteria in swimming pool water are made safe and pollution is removed expeditiously by circulating the water through an attached treatment plant. Swimming pool design is as important as chemical treatment in maintaining water quality, hence circulation systems normally include the ability to draw water from the surface as well as from the bottom of a pool. This ensures that surface pollution such as dust, floating debris (including grass in outdoor pools), hair, skin scales, body grease and mucus excreted from nose and mouth, are removed from the top layers while fluff, soil, sand from filters and precipitated chemicals are removed from the bottom. Dissolved pollutants such as suntan lotion, cosmetics, sweat and urine, and suspended materials such as sticking plasters and treatment chemicals from the middle layers of a pool are also removed by the appropriate positioning of water outlets.

The presence of a sufficient number of water outlets ensures that the outlet velocity is so low that no person or item is sucked against an outlet grill, and a sufficient number of inlets ensures that the water inlet velocity is so low that small children are not swept into deeper water. Pool water circulation pumps, with protecting grills in case debris reaches them, ensure an adequate turnover of water, ideally 30 litres of water per bather per day. This is achieved through a closed loop system from the pool, with filters to remove particulate matter, a disinfection plant and a chemical dosing plant that maintain a residual disinfectant and a chemical balance in the pool water, and finally a water heater. The disinfection and chemical treatments are designed to minimize the risks to pool users from the effects of skin, throat and faecal bacteria, saliva, nitrogenous matter from sweat and urine from fellow bathers, and the fungi and algae that could thrive in the water of a poorly maintained pool.

Even at the highest recommended concentrations, the chlorine-based disinfectants used in pool water rarely cause eye irritation. However, reactions between the disinfectant and urea (from sweat and urine) produce chloramines in the water, potentially contributing towards eye and skin irritation; in an indoor pool, these may also escape into the atmosphere over the water, affecting both staff and bathers. For this reason, bathers are encouraged to shower before entering a pool, thereby reducing the costs of chemicals to combat the urea and so minimizing chloramine irritation. Up to two-thirds of sweat products and a third of the bacteria are removed by showering, an effect that is, interestingly, not improved by using soap. An alternative method of disinfection by oxidation is ozonolysis (see Section 6.2.5), but obviously this can be considered only at the design stage for a swimming pool.

Legionellosis (Legionnaires' disease), a severe form of pneumonia caused by the *Legionella pneumophilia* bacterium, occurs only at poorly managed spa pools where the bacterium can spread in the fine spray generated at turbulent water surfaces. Supply water stored above 60 °C and piped at 50 °C or more kills this bacterium. For features at any pool with spray effects, such as fountains and rain sprays, bacterial levels are reduced by periodically flushing the spray heads with a solution of hypochlorite (OCl^-) at a concentration of 5–10 mg l^{-1}.

It is important that floor-cleaning materials and chemicals are not allowed to enter swimming pool water, as they may interact adversely with the treatment chemicals, as well as with bathers. However, some designs of pool surround make this difficult. If there is a likelihood of cleaning solutions entering a pool, the surrounds are scrubbed with pool water only. The floor and walls of the pool itself are brushed or vacuum-cleaned regularly to remove algae and debris.

Viscosity is a quality of swimming pool water not normally considered by the average bather. However, viscosity decreases with increasing temperature, by about 20% from 20 °C to 30 °C. The 1984 Olympic Games were held in very warm Los Angeles, in an outdoor pool, so the water was cooled with blocks of ice to prevent that year's swimmers claiming record times compared with other years when the events had been held in cooler climates.

6.1.3 Water for cooling industrial processes

Power stations and other heavy industrial users require very large volumes of water. The water is often drawn directly from rivers, estuaries or oceans and often contains considerable amounts of suspended and dissolved materials (Figure 6.1).

Figure 6.1 An electricity power station.

At a 2000 MW power station, up to 205×10^3 m^3 of water per day (45 million gallons) may be needed for cooling, although this is returned to the environment. The water quality may have to be modified by removing debris and larger suspended matter to prevent damage to, and blockages of, pipework, pumps, and balance and storage tanks. Such removal is achieved by installing coarse mesh screens or grills at the point where the water is drawn from the ocean or river. Finer screens are usually fitted behind these. To prevent the inlet becoming blocked, manual or automatic cleaning has to be carried out regularly. The larger debris such as trees, cut timber or carcasses are removed by hand from the coarse grills, but the design of the finer screens usually enables them to be cleaned automatically.

With a screen installed in a pit behind the coarse grills, the water passes from the outside through the periphery of the revolving screen to a pump suction on the inside, leaving the debris on the outside. As the screen rotates, the debris is lifted out of the water and removed by high-pressure water jets. Another design is the cylindrical screen consisting of a spiked cylinder revolving in the path of the incoming water. The debris is lifted out of the water on the spikes and removed, also by high-pressure water.

The expense involved in further treating cooling water by filtration is not generally justified, but the water will still contain organisms, which may settle in the pipework, restrict the water flow and adversely affect the heat transfer properties. In the condenser of a power station turbine, any loss of condensing capability will have serious implications for turbine efficiency and overall electricity production. Furthermore, once these organisms have settled within a cooling water system they are likely to grow, and may completely block the flow. Chlorine, either as a solution of chlorine gas in water or as sodium hypochlorite, is usually added to the cooling water after the screens to kill the organisms. The chlorine dose is carefully calculated so that there is just sufficient to fulfil its biocidal function within the cooling system but there is none (or a known minimum concentration) remaining in the water that is returned to the river or sea.

Figure 6.2 Cooling towers at a power station.

If cooling water is drawn from a restricted source, such as a slowly moving river, or during periods of drought, cooling towers may be incorporated. These can be of the convection (natural-draught) or forced-draught type. The most conspicuous cooling towers used in the UK are the huge hyperbolic concrete structures 115 metres or more high (Figure 6.2). In one design the water is pumped to the centre of the tower and allowed to fall over wooden, plastic or concrete baffles; as the water is broken into droplets, air is drawn in by convection currents and the water temperature is reduced by about 10 °C. In considerably smaller forced-draught towers the air is blown in or drawn in by huge fans. High towers eliminate air recirculation and so improve the cooling effect, and also disperse from the immediate surroundings any water droplets and vapour swept out by the convection currents (Figure 6.3). The same mass of water is, therefore, recirculated through the cooling circuit in the factory or power plant, although a certain amount (up to 25%) will be needed to make up losses.

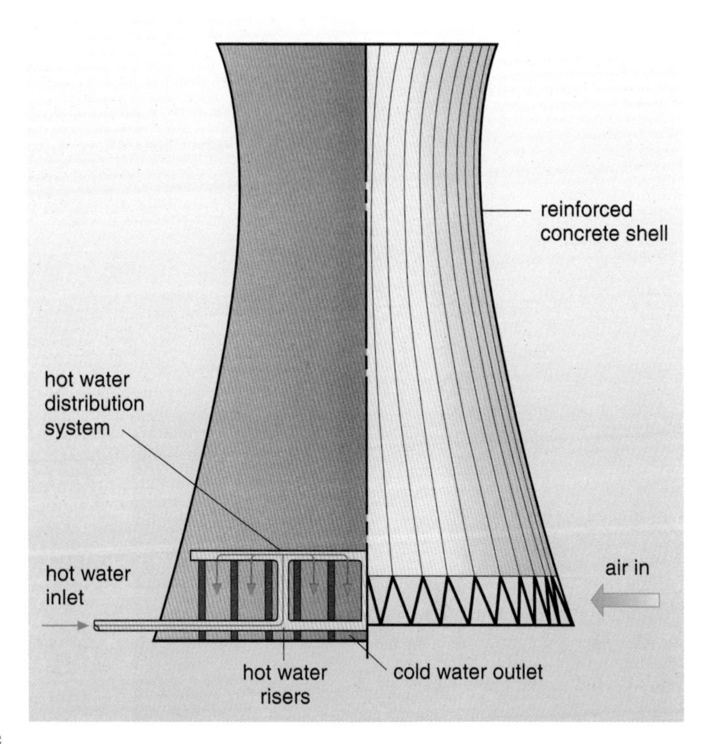

Figure 6.3 Diagram of a cooling tower showing a cross-section on the left.

Depending on its quality with regard to suspended or dissolved substances, cooling water can cause either thinning of the heat transfer surfaces (the cooling water pipes or condenser tubes) by erosion or corrosion, leading ultimately to their failure, or thickening of the heat transfer surfaces by the accumulation of sediment. In the first instance, a failed cooling pipe could introduce corrosive sea or river water into otherwise pure water or steam. In the second instance, thickening of a heat transfer surface will lead to a fall in cooling efficiency and, in a power plant, even greater detrimental effects, such as the loss of vacuum in a turbine. As the huge volumes of water that are used for cooling do not merit filtration or removal of the dissolved solutes, damage by erosion and/or corrosion and also sediment accumulation have to be minimized by such measures as: using more resistant coolant pipes (install expensive hard titanium instead of cheaper soft copper, for example), employing better pipework design (no sharp bends causing drastic changes in direction or velocity of water flow) and regular removal of sediment. Specialized coolers, such as those in large electrical generators, require the cooling water to be extremely pure. In this situation, demineralized water is used.

6.2 Methods of water treatment

6.2.1 Desalination

Desalination, or the removal of common salt (sodium chloride) or, indeed, other salts from a saline or salt solution, is performed selectively by vegetation all the time. The extraction of freshwater from saltwater by evaporation or distillation has been undertaken on ships for almost 200 years.

Methods of desalination fall into two categories: removing water from the salt, or removing salt from the saline water. In the first category are evaporation or distillation, reverse osmosis, freezing and hydrate adsorption; in the second are electrodialysis, osmionic processes, thermal diffusion and ion exchange. We shall explore some examples later.

6.2.2 Natural purification

Exposure to both sunlight and air greatly helps to purify water naturally. The ultraviolet rays in sunlight are effective in killing bacteria in water and, in sunlight, aquatic plants and algae are able to photosynthesize, thereby reducing the amount of carbon dioxide and increasing the amount of oxygen in the immediate vicinity.

Aeration is another natural method of water purification. As water breaks into droplets, especially at waterfalls, where a fine spray is created, or where the wind can whip up the surface of a river or lake, water necessarily becomes aerated (Figure 6.4).

Figure 6.4 Water being aerated in a waterfall.

The volume of river water flowing past a particular point is the same as that flowing past any other point (unless streams or other sources join the river, adding to the flow). The cross-section of a stream or river, however, varies throughout its length. At some places, particularly in the upper reaches where a river narrows, its cross-section shrinks, and the water velocity increases; while at other places, particularly in the lower reaches where the river widens out, the cross-section increases and the water velocity is considerably lower. When water is flowing quickly, the suspended matter that we mentioned above tends to remain in suspension but where the water flow is slow, the matter in suspension drops through the water to the river bed, settles and accumulates as a sediment.

Clarification through sedimentation is another important natural process of purification. Some organisms will be deposited in the slow reaches of a stream or river and may die and decay. Any tiny particles of mineral matter in suspension are also likely to be deposited, forming a sediment with the organic materials. Anaerobic bacteria in the mud remove oxygen from sulfates, release hydrogen sulfide and so create the characteristic smell of rotten eggs in mud found in harbours or estuaries. This natural process of sedimentation has been copied with very effective results in artificial methods of purification.

When streams or rivers are in flood, the water erodes the banks and becomes clouded by the materials that are swept off. If floodwaters are fairly gentle in their action, a change in colour, rather than turbidity, may occur. Both coloration and turbidity are due to matter in suspension, and disappear when water flows so slowly that the particles settle to the river bed. If the velocity is high, particles may not immediately sink to the bottom, but may be carried on to lower reaches, forming mud and sandbanks which, in time, may change the course of the river.

So, a flowing stream or river heavily charged with impurities is purified by nature if it is allowed to flow long enough over an uncontaminated stretch of river bed. If, however, effluents containing certain wastes are discharged into the stream or river, purification by nature alone may not be sufficient and artificial means may have to be introduced.

6.2.3 Artificial purification

Clarification, aeration and slow sand filtration

Water that is held in an artificial reservoir for at least two months loses much of its suspended matter and is clarified by simple sedimentation. In reservoirs that have a large surface area, there is an opportunity for wind to whip up the surface of the water causing a little aeration, so reproducing the natural purification process described above. In some reservoirs in the USA, the water is pumped into the air as a fine spray to introduce oxygen artificially.

Following clarification in large reservoirs, the water is usually filtered by **in-depth filtration** through a filter bed. Sand, (or anthracite or marble) filter beds are used when the quantity of suspended solids is large but their particle size is relatively small. The solids must be able to penetrate the bed to a sufficient depth yet not clog its surface. The other basic method of filtration, normally for smaller volumes of water and for removal of finer particles, is by **surface filtration**, through various types of supported porous media. Examples of surface filters include strainers and meshes of macro (above 150 μm) and micro (less than 150 μm) sizes, replaceable cartridges, sintered materials and perforated metal or plastic cylinders (termed candles) coated with cellulose or diatomaceous earth.

In sand filter beds the reservoir water falls slowly under gravity through clean sharp sand that contains minute particles of quartz. The ideal filter rate is about 10 cm per hour, yielding over 2000 litres per day through one square metre of sand. The space over the filter bed is designed to hold a volume of water to create the appropriate head of water for an optimal filter rate. These conditions change, of course, as the filter begins to work and retain any suspended matter that has not been removed by sedimentation.

The retained impurities initially assist in the filtration process, but gradually slow down the flow rate and would eventually clog up the filter bed if left unattended. Initially the water head is increased to maintain an acceptable filtration rate, but ultimately the filters have to be cleaned. This is normally effected either by scraping off the top layer of sand, which is subsequently washed before being put back in place, or by backwashing the whole filter bed unit – that is water is pumped 'backwards' up through the sand so washing off any intercepted matter through suitable pipework to waste. In situations where water has been clarified before filtration, slow sand filters can last for several months before being cleaned. However, in situations where water is taken directly from a river, the filters may require cleaning every few weeks.

Slow sand filtration is extremely efficient and requires little maintenance. It was formerly the preferred method of filtration almost universally, as it removes all suspended and colloidal matter and water coloration caused by particulate matter in suspension. However, these filter beds occupy large areas and they are unable to remove solutes from water. Some solutes, when present at very low levels pose no health hazard, but cause water coloration. Nonetheless coloured tap water is unacceptable in many parts of today's world. Hence slow sand in-depth filtration has been replaced in these places by fast sand in-depth filtration.

Fast sand filtration and coagulation

Sharp sand is used as the filter medium, as in slow sand filtration, but in this case the sand is held in considerably smaller filter beds, normally contained in cylindrical tanks. Reservoir water is pumped through the filter beds at about 5 times the rate of a slow filter, in other words at about 0.5 m an hour (or about 12 000 litres a day) per square metre of filter surface. To enable fast sand filtration to be more efficient, the clarification process in the reservoir prior to filtration is enhanced by adding a coagulant or a flocculating agent to the water. This forms blancmange-like flocs with solutes, including those that are coloured. As the flocs sink through the reservoir water, they entrap other suspended and colloidal materials, including bacteria, and carry them to the bottom. The clarified water is drawn from above the sediment.

Chemicals used as coagulants include aluminium sulfate (alum), polyaluminium chloride (PAC) or aluminium hydroxychloride, polyaluminium sulfo-silicate (PASS) and sodium aluminate, all of which work well at pH values between 6.5 and 7.2:

$$\underset{\text{alum}}{Al_2(SO_4)_3} + \underset{\text{hardness}}{3Ca(HCO_3)_2} = \underset{\text{floc}}{2Al(OH)_3} + 3CaSO_4 + 6CO_2$$

$$\underset{\text{sodium aluminate}}{NaAlO_2} + \underset{\text{hardness}}{Ca(HCO_3)_2} + H_2O = \underset{\text{floc}}{Al(OH)_3} + CaCO_3 + NaHCO_3$$

Sodium aluminate also produces a floc with dissolved carbon dioxide:

$$\underset{\text{sodium aluminate}}{2NaAlO_2} + \underset{\text{carbon dioxide}}{2CO_2} + 4H_2O = \underset{\text{floc}}{2Al(OH)_3} + 2NaHCO_3$$

Iron(III) chloride and iron(III) sulfate have also been used as coagulating agents, but obviously increase the iron concentrations, and may cause colouring and increase the turbidity of the water, these compounds work well between pH 6.5 and 7.5:

$$\underset{\text{iron(III) chloride}}{2FeCl_3} + \underset{\text{hardness}}{3Ca(HCO_3)_2} = 3CaCl_2 + \underset{\text{floc}}{2Fe(OH)_3} + 6CO_2$$

$$\underset{\text{iron(III) sulfate}}{Fe_2(SO_4)_3} + \underset{\text{hardness}}{3Ca(HCO_3)_2} = 3CaSO_4 + \underset{\text{floc}}{2Fe(OH)_3} + 6CO_2$$

The addition of any treatment chemical has to be carefully considered and all side-effects estimated to establish whether the treatment is detrimental to, or incompatible with, subsequent storage, treatment and distribution systems. The overall objective is the provision of water that is fit for human consumption.

Following flocculation and sedimentation, the water is passed to fast in-depth sand filters. To maintain the filter beds in optimum condition they are regularly backwashed.

The filtered water is now clear, but may require further chemical treatment. The addition of an alkali may now be necessary if an added coagulant decreased the pH. Lime ($Ca(OH)_2$) is often used to restore the pH, particularly if the raw water is very soft and domestic distribution systems are known to contain lead:

$$CO_2 + Ca(OH)_2 = CaCO_3 + H_2O$$

If the lime reacts with the aluminium sulfate directly, the following reaction takes place:

$$Al_2(SO_4)_3 + 3Ca(OH)_2 = 2Al(OH)_3 + 3CaSO_4$$

Whilst the chemical coagulation and backwashing processes involve operating costs that slow sand filters do not incur, fast sand filtration is the preferred technique in many parts of the world. The smaller areas occupied by the filter beds is an essential financial consideration in many countries, especially if treatment plants are near cities with high land prices.

Reverse osmosis

This process (also termed nanofiltration and hyperfiltration) utilizes identical principles, but reversed, to natural osmosis. Osmosis is a particular example of diffusion of a solution, but through a semipermeable membrane where the pore sizes permit only water molecules to pass through the membrane. In reverse osmosis, instead of osmotic pressure being responsible for the passage of water through a membrane to a region of higher concentration, pressure is artificially applied over water that contains solutes as unwanted impurities. Only water molecules, and other substances of the same or smaller sizes, can pass through the pores of the semipermeable membrane, and so the water is purified by what is, in effect, filtration at a very fine level.

The semipermeable membranes used in reverse osmosis are made of synthetic materials. Today the membranes are usually made in long cylinders, rather like large macaroni. In some designs, very high pressure is imposed on impure water on the outsides of the membranes and pure water is withdrawn from the insides; in other designs the flow is in the opposite direction (Figure 6.5). When the membranes become blocked, as in the case of sand filters, they can either be similarly backwashed or replaced. The huge advantage of reverse osmosis in water purification is that no treatment chemicals are involved and the results are impressive (Table 6.3).

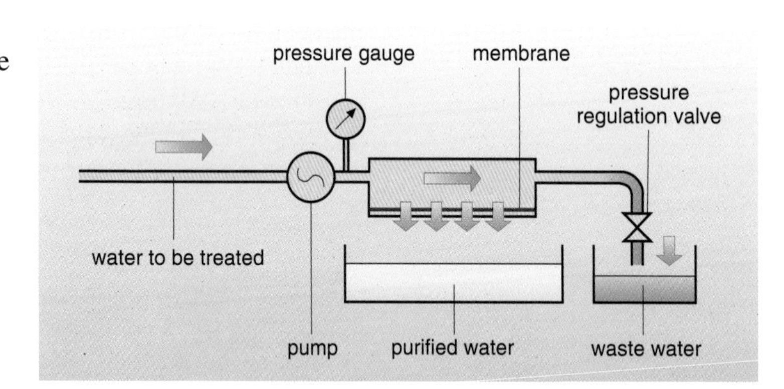

Figure 6.5 Diagram of a reverse osmosis system.

Table 6.3 Typical values of water quality parameters before and after treatment by reverse osmosis. (Concentrations of substances are in mg l^{-1}.)

Parameter	Raw water concentration	Treated water concentration
Ca^{2+}	146	5
Mg^{2+}	42	1
Na^{+}	43	28
SO_4^{2-}	171	5
Cl^{-}	58	24
HCO_3^{-}	370	80
NO_3^{-}	0.4	0.1
TDS	830	150
pH	7.05	6.0

Question 6.1

Given that reverse osmosis is very effective in reducing the concentrations of dissolved ions in water, why is it not generally used in the treatment of water?

Evaporation

The hydrological cycle involves water vapour leaving the oceans by evaporation and forming clouds in the atmosphere. Various atmospheric gases and particulate matter of all kinds can become dissolved and entrapped in the condensation of this otherwise pure form of water. If water vapour is cooled under controlled conditions, as in water distillation equipment in laboratories, the chance of dissolution or entrapment of impurities in the collected condensate or distillate is minimal. Similarly prepared distilled water is used for clinical and industrial purposes worldwide where water with high purity is required.

Distillation, i.e. evaporation and subsequent condensation, is not a new idea. Many thousands of years ago, peoples in coastal deserts built condensation cairns from local stone. Damp breezes circulated naturally through passages in the cairns, and the thick cooler stones on the inside, on which condensation took place, were angled to enable the dripping condensate to be collected. Tests have shown that such techniques were not insignificant. A sea breeze at 30 °C with 80% relative humidity contains about 25 g of water per cubic metre. If this air is cooled to 10 °C it yields around 15 g of water. If the breeze is blowing at 5 m s^{-1}, the condensate will form on a square metre of cool surface at a rate of 75 g every second or 4.5 litres per minute. In a four-hour night-time cooling period this amounts to 1080 litres of 'pure' water. Obviously, if the cooling area was to be increased, the volume of condensate could be considerable. Some desert plants acquire their water by this method.

On a large commercial scale, water is purified in evaporators and then condensed under controlled conditions that prevent the dissolution or inclusion of other substances. For example, solutes can become entrained in the vapour during boiling if the heating is too vigorous. To overcome this, it is normal to install two or three evaporator units in series. At some locations, where the raw water is hard, a softening unit is fitted before the evaporator so that the residue in the evaporator becomes a soft sludge instead of a hard scale.

Evaporation plants require only a method of heating raw water to its boiling temperature and then condensing the pure vapour generated. In addition, storage facilities are required for the distilled water, as well as a means of removing the solutes and suspended substances that remain in the boiler. The heat can come from burning coal, oil or natural gas, or the fission process of a nuclear reactor. Any of these sources could technically be associated with the generation of electricity if designed appropriately, but obviously economic considerations have to be taken into account.

6.2.4 Hardness reduction and removal

Softening water

Substances that are regarded as hardness salts contain calcium and magnesium ions. When water is softened, these ions are replaced by sodium ions. Hardness is measured as the concentration of calcium or magnesium ions in the water, but it is customary for industrial water chemists to express the concentration of all hardness salts as parts of calcium carbonate per million parts of water by mass, or ppmm $CaCO_3$. For solutes in water this is essentially the same as parts per million by volume. Research chemists tend to use the units of micrograms per millilitre ($\mu g\ ml^{-1}$) or milligrams per litre ($mg\ l^{-1}$).

Rainwater has a hardness of about 5 ppm or less, some river waters about 50 ppm, but in chalky areas the river hardness may be 400 ppm $CaCO_3$.

Cold lime–soda softening

Lime (calcium hydroxide) is added to hard water at ambient temperatures. This causes several hardness salts dissolved in the water to be precipitated as insoluble materials. The hardness salts are thereby effectively removed from the water, falling to the bottom of the vessel:

hardness		lime		precipitate				solution
$Mg(HCO_3)_2$	+	$2Ca(OH)_2$	=	$Mg(OH)_2$	+	$2CaCO_3$	+	$2H_2O$
$MgSO_4$	+	$Ca(OH)_2$	=	$Mg(OH)_2$			+	$CaSO_4$
$Ca(HCO_3)_2$	+	$Ca(OH)_2$	=	$2CaCO_3$			+	$2H_2O$
$MgCl_2$	+	$Ca(OH)_2$	=	$Mg(OH)_2$			+	$CaCl_2$

Soda ash (the old, industrial name for sodium carbonate, Na_2CO_3) is then added to the solution containing the calcium salts to convert them to the corresponding sodium salts and precipitate calcium salts out of solution:

solution		soda ash		precipitate		soft water
$CaSO_4$	+	Na_2CO_3	=	$CaCO_3$	+	Na_2SO_4
$CaCl_2$	+	Na_2CO_3	=	$CaCO_3$	+	$2NaCl$

These processes are carried out in large tanks, the lime and soda being continuously added as the hard water flows in while the entire contents of the tank are agitated mechanically; the precipitate is drawn off from the bottom, the softened water from the top of the tank. This process is used to treat water having a hardness of 300 ppm or more, and reduces the concentration to about 40 ppm. The hardness cannot be reduced further because at ambient temperatures the reactions between the hardness salts and the treatment chemicals are slow. Furthermore, calcium carbonate and magnesium hydroxide are, in fact, not completely precipitated and some of these salts remain in solution.

Sodium carbonate has been added to soften water for well over a century in the domestic situation, but with slightly different terminology. Washing soda crystals (the old colloquial name for sodium carbonate crystals) were sprinkled by our great-grandparents and their grandparents into their scullery washing-up water to reduce the effects of temporary and permanent hardness before washing their crockery and cutlery. Bath salts also originally comprised only sodium carbonate crystals with the absolute minimum of scent and colouring. In fact, in many households, simple washing soda was used in bath water. However, washing soda creates quite alkaline conditions and its use has largely been phased out as people have become more concerned over skin care.

temporary hardness	+	washing soda	=	precipitate	+	soft water
$Ca(HCO_3)_2$	+	Na_2CO_3	=	$CaCO_3$	+	$2NaHCO_3$
permanent hardness	+	washing soda	=	precipitate	+	soft water
$MgSO_4$	+	Na_2CO_3	=	$MgCO_3$	+	Na_2SO_4

Other sodium salts, sodium metaphosphate and sodium sesquicarbonate, have generally replaced sodium carbonate in household use.

Hot lime–soda softening

Exactly the same chemicals are used as in the cold process, but superheated steam is sprayed in with the hard water, raising the operating temperature to about 120 °C. This enables the chemical reactions to take place almost instantaneously and calcium carbonate and magnesium hydroxide are more completely precipitated. Less treatment chemicals are required and the hardness is reduced to about 10 ppm. A filter is usually incorporated after this treatment before the water is used.

Lime–zeolite softening

Zeolites are naturally occurring materials with the capability of exchanging ions. The name *zeolite* is also applied to synthetic resins that have been especially manufactured for this purpose. In lime–zeolite softening, calcium hydroxide is added, as in the first stage of lime–soda softening, producing a solution containing calcium sulfate, calcium chloride. This solution is then filtered and treated with a sodium zeolite which exchanges its sodium ions for calcium ions, thus removing the calcium ions from solution:

solution	+	sodium zeolite	=	calcium zeolite	+	soft water
$CaSO_4$	+	Na_2Z	=	CaE	+	Na_2SO_4
$CaCl_2$	+	Na_2Z	=	CaE	+	2NaCl

No further filtration is required, and the water produced has virtually zero hardness.

The (sodium) zeolite can adsorb only a certain amount of calcium ions onto itself. When it cannot take any more, it is said to be 'exhausted'. Zeolites, however, can be 'regenerated', i.e. restored to their original state, by washing off the calcium ions with a more concentrated solution of sodium ions. Sodium chloride is a cheap and convenient chemical to use as a source of sodium ions. The waste solution is rejected:

exhausted zeolite	+	regenerant solution	=	regenerated zeolite	+	waste solution
CaZ	+	2NaCl	=	Na_2Z	+	$CaCl_2$

Cation exchange softening

For waters that have low hardness (around 20 ppm $CaCO_3$, for example), lime treatment is unnecessary and all hardness salts can be softened with a cation exchange resin, similar to the zeolite material described above.

hardness salts		cation exchanger		held on		soft water
$MgCO_3$	+	Na_2E	=	MgE	+	Na_2CO_3
$Ca(HCO_3)_2$	+	Na_2E	=	CaE	+	$2Na(HCO_3)_2$
$MgCl_2$	+	Na_2E	=	MgE	+	2NaCl
$CaSO_4$	+	Na_2E	=	CaE	+	Na_2SO_4
$CaCl_2$	+	Na_2E	=	CaE	+	2NaCl

This treatment produces water with zero (or extremely low) hardness. When exhausted, the exchange resin is regenerated with sodium chloride as in the lime–soda softening system.

Demineralizing water

This method of treatment is based on the same principles utilized in cation exchange softening, but the treatment is taken a stage further with both cations and anions exchanged. In the demineralization process, the cations are replaced with hydrogen ions instead of sodium ions, and the anions are replaced with hydroxide ions. If we assume that the water whose quality we wish to alter contains 100% sodium chloride, the following reactions will occur:

	salt		exchanger		held on exchanger		effluent
cation exchange:	NaCl	+	HE	=	NaE	+	HCl
anion exchange:	NaCl	+	XOH	=	XCl	+	NaOH

Notice that the effluent from the cation exchange reaction is acidic while from the anion exchange it is alkaline.

The materials used as exchangers commercially are usually synthetic resins on a polystyrene framework, and not naturally occurring materials; this enables a consistent product to be used and eliminates natural impurities that normally accompany greensands. In the equations above, the first resin (shown as HE for simplicity) is properly termed a cation exchange resin in the hydrogen form, or a hydrogen ion exchange resin. The second resin (XOH) is termed an anion exchange resin in the hydroxide (or hydroxyl) form, or a hydroxide (or hydroxyl) ion exchange resin.

If the two exchange reactions are undertaken sequentially whereby the effluent from the hydrogen ion exchange resin is passed directly into the hydroxide ion exchange resin, you can see that the effluent will be hydrogen hydroxide (water). If the exchange is complete, all the mineral salts or ions have been removed from the original raw water.

	salt		cation resin		held on resin		effluent	
cation exchange:	NaCl	+	HE	=	NaE	+	HCl	
	effluent from cation exchange		anion resin		held on resin		effluent	
anion exchange:	HCl	+	XOH	=	XCl	+	HOH	(H_2O)

As all of the mineral ions have been removed, this process is termed demineralization. Some water treatment companies (and even a few academics) refer to this process as deionization and the end product as deionized water, but you can see that this is not correct. The water has not had all of its ions taken away, they have only been replaced with others. (Even pure water contains small concentrations of H^+ and OH^- ions.)

As in lime–zeolite softening and in cation exchange softening, when all the ion exchange sites have fulfilled their tasks, the resins are termed exhausted. Likewise, the exhausted resins can be regenerated, but in this instance the replacement ions needed are hydrogen and hydroxide. Ideal sources of these are highly ionized acids and alkalis such as sulfuric acid and sodium hydroxide, although the actual choice will be a matter of economics and convenience. In the case of our example the chemical reactions involved are:

	exhausted resin		regenerant		regenerated resin		waste liquors
cation resin:	2NaE	+	H_2SO_4	=	2HE	+	Na_2SO_4
anion resin:	XCl	+	NaOH	=	XOH	+	NaCl

A vessel containing a homogeneous mixture of cation and anion exchange resins, termed a mixed-bed (MB) unit, produces water of greater purity than the two-stage demineralization arrangement described above. Regeneration of a mixed-bed unit is similar, but the resin mixture has first to be separated into two layers, one solely of cation resin, the other of anion resin. This is achieved quite easily by backwashing the resin mixture. Water is pumped in at the bottom of the unit and out of the top (in the opposite direction to the normal flow); the cation resin, being denser than the anion resin, separates as the lower layer. Acid (hydrogen ions) is then passed through the bottom resin while alkali (hydroxide ions) is passed through the top layer, thereby reducing regeneration time. The regenerated resins are subsequently intimately mixed by blowing air in at the bottom of the unit, which is then filled with water and returned to service.

Other combinations of ion exchange units are installed depending on the quality of the raw water and the required quality of the treated water. For example, raw water with low chloride and sulfate concentrations but relatively high carbonate and bicarbonate content can be effectively demineralized. The water passes through a cation exchange resin in the hydrogen form and the released carbon dioxide is physically removed by blowing air upwards over falling effluent to sweep out the carbon dioxide with air. Then, a mixed-bed unit is used to remove any unwanted ions that have slipped through the earlier treatment.

raw water		exchanger		held on exchanger		effluent		
Na_2CO_3	+	2HE	=	2NaE	+	H_2O	+	CO_2

6.2.5 Water sterilization

Chlorination

Chlorine gas dissolves in water forming a solution containing hypochlorous and hydrochloric acids in a reversible reaction:

chlorine		water		hypochlorous acid		hydrochloric acid
Cl_2	+	H_2O	=	HOCl	+	HCl

Hypochlorous acid is a most effective agent in the sterilization and disinfection of water because it destroys potentially harmful microbes. Chlorine gas is not easily stored or handled and so solid hypochlorites are often used in preference at a water treatment plant. Sodium and calcium hypochlorites in water also produce hypochlorous acid, the main reactions being:

sodium hypochlorite		water		hypochlorous acid		sodium ion		hydroxide ion
NaOCl	+	H_2O	=	HOCl	+	Na^+	+	OH^-

calcium hypochlorite		water		hypochlorous acid		calcium ion		hydroxide ion
$Ca(OCl)_2$	+	$2H_2O$	=	2HOCl	+	Ca^{2+}		$2OH^-$

From the equations above you can see a major distinction in that chlorine gas in water reduces the pH whereas both sodium and calcium hypochlorite increase the pH. However, the situation is a little more complicated than this because hypochlorous acid is a weak acid but dissociates to some extent into hydrogen and hypochlorite ions:

hypochlorous acid		hydrogen ion		hypochlorite ion
HOCl	=	H^+	+	OCl^-

Hypochlorous acid and hypochlorite ion together provide the free residual chlorine in a particular water, although the hypochlorous acid is the better disinfectant.

As noted above, hypochlorous acid is a weak acid: if a solution of the acid is analysed, then we find that some, *but not all*, of the HOCl is dissociated into H^+ and OCl^- ions. Moreover, if we could see what was happening at the molecular level, we would find that the *rate* at which HOCl is dissociating into H^+ and OCl^- is exactly matched by the rate at which these ions are recombining to form HOCl. The solution is an example of a system in a state of *dynamic* equilibrium. And such systems respond to being disturbed in a characteristic way, encapsulated by Le Chatelier's principle: put simply 'for a system in equilibrium, if any change is made to the conditions, the system will respond so as to oppose the change'. In short, it's as if the system has an in-built negative feedback mechanism.

To return to a solution of HOCl, if acid is added to the solution, the concentration of H^+ ions is increased (lowering the pH), and this disturbs the equilibrium. The system responds by 'using up' the additional H^+ ions: they combine with OCl^- ions to form HOCl. The upshot is that the concentration of hypochlorous acid increases, and that of hypochlorite decreases. As it is the hypochlorous acid that is the more effective agent in disinfection, the lower the pH, the greater the disinfectant effect. Conversely, the higher the pH (the greater the alkalinity) the lower the disinfection even though the overall free chlorine level is maintained. The chemical mechanism by which chlorine acts as a disinfectant is as follows: unwelcome pollutants such as ammonia are generated by bacterial activity on nitrogenous compounds in water; chlorine reacts with ammonia to produce monochloramine:

chlorine		ammonia		hydrochloric acid		monochloramine
Cl_2	+	NH_3	=	HCl	+	NH_2Cl

The monochloramine that is produced also has some disinfection properties but is not as effective as free chlorine. It is also not an irritant. However, if this monochloramine reacts with more chlorine, the irritant dichloramine is produced:

chlorine		monochloramine		hydrochloric acid		dichloramine
Cl_2	+	NH_2Cl	=	HCl	+	$NHCl_2$

Further chlorination forms trichloramine (also called nitrogen trichloride) which is an even greater irritant, especially for sensitive eyes, and has a characteristic smell (usually attributed to chlorine). This can be of concern when treating swimming pool water directly, compared with disinfecting drinking water at a water company's premises.

chlorine gas		dichloramine		hydrochloric acid		trichloramine
Cl_2	+	$NHCl_2$	=	HCl	+	NCl_3

On the other hand dichloramine can combine with monochloramine to produce innocuous nitrogen (and hydrochloric acid which will further reduce the pH).

monochloramine		dichloramine		hydrochloric acid		nitrogen
NH_2Cl	+	$NHCl_2$	=	3HCl	+	N_2

The important general point is that dosing water with chlorine requires careful consideration of pH and calculation of pollutant concentrations to achieve the optimum disinfection.

Ozonolysis

Ozone is an effective bactericide but it does not react with ammonia, produced by bacterial attack and decay. Nevertheless, it is an extremely reactive oxidant and is toxic in high concentrations. Ozone levels of $10\,mg\,m^{-3}$ in air can lead to skin irritation and breathing difficulties. It is unstable and relatively insoluble in water, so has to be generated and used *in situ* at a water treatment plant.

Ozone is generated by electrical discharge produced by an A.C. voltage of around 12 000 volts. Air is dried and filtered to remove dust particles before being subject to the electric discharge. Some of the oxygen in the air is converted to ozone, in a similar way to atmospheric oxygen being converted to ozone during an electric thunderstorm. It requires about 20 watts of power to produce a gram of ozone.

6.2.6 Water fluoridation

In areas with a naturally high fluoride ion concentration in drinking water, there are fewer cases of dental caries (tooth decay). The breakdown of calcium hydroxyapatite, $Ca_5(PO_4)_3OH$, the main constituent of tooth enamel, is another example of an equilibrium process:

$$Ca_5(PO_4)_3OH = 5Ca^{2+} + 3PO_4^{3-} + OH^-$$

This reaction could be moved to the right (enamel breakdown increased) by removing hydroxide ions. In the mouth, enzymes from bacteria in plaque (a gelatinous collection of microbes) catalyse the fermentation of sugars into lactic acid which forms lactate salts with hydroxide ions:

$$\text{Lactic acid} + OH^- \rightarrow \text{Lactate} + H_2O$$

Fluoride ions inhibit the enzymes and also form some fluoridated hydroxyapatite, $Ca_5(PO_4)_3OH_{1-n}F_n$, in the enamel, which is less susceptible to attack by acid than the original hydroxyapatite. However, very high fluoride levels for prolonged periods can damage kidney, thyroid and bone, but up to 1 ppmm fluoride in drinking water appears to offer only beneficial water effects. Many water supplies do not contain fluoride naturally at this concentration and some water companies add it to benefit their customers.

6.2.7 Oxygen removal

We said earlier that while aquatic organisms require dissolved oxygen in water for their survival, many industrial systems fabricated from metals in which aqueous solutions are circulated are likely to become oxidized, corroded and eventually fail if oxygen is present. Depending on the system and its design, air, with its oxygen, can be removed from water mechanically or physically by agitation and creating a vacuum over the water so that the air is effectively ‘pulled out’ of the water. Industrial de-aerators operate on this principle.

Oxygen can also be removed chemically by adding a reducing agent such as sodium sulfite or hydrazine to the water:

sodium sulfite	+	oxygen	=	sodium sulfate		
$2Na_2SO_3$	+	O_2	=	$2Na_2SO_4$		
hydrazine	+	oxygen	=	nitrogen	+	water
N_2H_4	+	O_2	=	N_2	+	H_2O

Hydrazine has the advantage of being a liquid and the products of oxygen removal are nitrogen (chemically inert) and water. Although solid sodium sulfite is cheaper than hydrazine and dissolves without difficulty in water, the end-product of the reduction reaction is an aqueous solution of sodium sulfate, which noticeably increases the total dissolved solids (TDS) content of the water.

6.2.8 Corrosion inhibition

To minimize corrosion in relatively small-volume closed water circuits, such as central heating systems in homes, hotels and offices, a chemical or chemicals are added to the circulating water to form a protective coating on the internal metal surfaces and so inhibit further chemical attack. Long-chain fatty amines can be used to form continuous films, in effect waterproofing the metal surfaces. Soluble oils offer similar protection. The success of this protection depends on the films being continuous because corrosion is enhanced at any spots of exposed metal. Inorganic inhibitors such as chromates, polyphosphates and silicates have also been used. Inorganic inhibitors initially cause a small amount of corrosion to the metal surfaces but the corrosion products adhere to the metal and form a protective layer that inhibits further corrosion.

6.3 Summary of Section 6

1. There is a range of statutory standards for drinking water and for water for other uses. For some industrial uses, drinking water standards are insufficiently stringent and water has to be further purified.
2. Water is purified by a number of natural processes that include sedimentation and aerial oxidation.
3. Artificial purification methods include sedimentation, filtration, coagulation, reverse osmosis and evaporation.
4. Domestic water supplies are sterilized and, in some cases, hardness is reduced.

Learning outcomes for Topic 6

1.1 List the uses to which water can be put and indicate 'quality' requirements for each.

1.2 Indicate why it is not useful to grade water from low to high on a single quality scale.

2.1 Calculate the effects of changing the temperature of a solution on the concentration and potential loss of a solute (solid or gas).

2.2 Indicate why a measurement of electrical conductivity gives information about dissolved materials.

3.1 Identify differences, in terms of suspended and dissolved materials, in water from different sources.

3.2 Explain the difference between permanent and temporary hardness and indicate the circumstances in which water is likely to be hard.

4.1 Outline major distribution methods for water and indicate criteria that need to be satisfied for specific outcomes.

5.1 Outline sources of contamination of water and indicate the contaminants specific to each source.

5.2 Explain how biological oxygen demand can be used as a measure of organic contamination of water.

5.3 Use the HERMES model to follow the progress of a non-conservative pollutant such as biological oxygen demand along a river.

6.1 Explain why different standards are applied to water according to intended use and give examples.

6.2 Outline methods of water treatment and suggest appropriate methods for specific purposes.

6.3 Indicate why and how cooling water from power stations is treated.

Answers to questions

Question 2.1

From Figure 2.1, the solubility of sodium nitrate is about 93 g per 100 g of water at 30 °C. Cooling the solution to 30 °C would result in sodium nitrate coming out of solution. For 110 g of the solution, (110–93)g sodium nitrate would be deposited. As we started with 1.5 l of solution, the quantity of sodium nitrate deposited is:

$$17\text{ g} \times \frac{1500}{100} = 255\ g$$

(assuming that 1 l of solution has a mass of 1000 g).

Question 3.1

As a result of ion exchange at the leaf surface, rain intercepted by forest is likely to have a higher pH than water that falls directly on the ground. Hydrogen ions in the raindrop are replaced by metal ions adsorbed on the surface of the leaf. The less acidic water will be less effective at dissolving the calcium carbonate and should result in less hard water.

Question 6.1

Like the use of many processes, the use of reverse osmosis is a compromise between technical efficiency, output and cost. The process is much more costly in comparison with many other methods of purification even though it is effective. Costs are linked to capital equipment, membrane renewal and the cost of pressure operation. It is used where there is a need for water of a particular quality.

Acknowledgements for Topic 6 *Water Quality*

Grateful acknowledgement is made to the following sources for permission to reproduce material in this book:

Figures 1.1, 1.3, 1.4a, b, 1.6, 1.7, 4.2, 5.1, 5.3, 6.4: Stuart Bennett, *Figure 1.2*: Courtesy of Salt Union Ltd, *Figure 1.8*: Robin and Arlene Karpan, *Figure 4.1*: Albert Berenguier, Jerrican / Science Photo Library, *Figure 6.1*: Chris Knapton / Science Photo Library, *Figure 6.2*: Sinclair Stammers / Science Photo Library.

Every effort has been made to trace all the copyright owners, but if any has been inadvertently overlooked, the publishers will be pleased to make the necessary arrangements at the first opportunity.

TOPIC 7

EUTROPHICATION

Joanna Treweek and David Gowing

Introduction

1

1.1 Definition

Eutrophication describes the biological effects of an increase in the concentration of nutrients. The collective term 'nutrients' refers to those elements that are essential for primary production by plants or other photosynthetic organisms. Eutrophication is most often caused by increases in the availability of nitrogen and phosphorus, commonly present in soil and water in the form of nitrate and phosphate, respectively. However, altered concentrations of any plant nutrient may have a recognizable biological effect. Eutrophication can occur in any aquatic system (freshwater or marine), and the term is also used to describe the process whereby terrestrial vegetation is affected by nutrient-enriched soil water.

1.2 Origin of the term 'eutrophication'

The levels of nutrients present determine the trophic state of a water body, where trophic means 'feeding'.

❍ Give another example of the adjective *trophic* being used in a scientific context.

● Trophic levels, as applied to a food chain.

The adjective *eutrophe* (literally 'well fed') was first used by the German botanist Weber in 1907, to describe the initially high nutrient conditions that occur in some types of ecosystem at the start of secondary succession. Scientists studying lakes at the beginning of the 20th century identified stages in plant community succession that appeared to be directly related to trophic state or nutrient status. They described a series of stages:

'oligotrophic — mesotrophic — eutrophic — hypertrophic'

where **oligotrophic** meant 'low in nutrients', **mesotrophic** 'with intermediate nutrient concentration', **eutrophic** 'high in nutrients' and **hypertrophic** 'very high in nutrients'. At the time, these definitions were derived from comparative estimates between water bodies with different nutrient status, judged according to their **phytoplankton** communities. Phytoplankton is a collective term for the free-floating photosynthetic organisms within the water column. It encompasses both algae (from the kingdom Protoctista) and photosynthetic members of the kingdom Bacteria. Thus an oligotrophic lake would have clear water with little phytoplankton, whereas a eutrophic lake would be more turbid and green from dense phytoplankton growth, and a mesotrophic lake would be intermediate between the two. Table 1.1 summarizes some of the general characteristics of oligotrophic and eutrophic lakes. A further definition, **dystrophic**, describes 'brown-water lakes', which have heavily stained water due to large amounts of organic matter usually leached from peat soils. The presence of these organic compounds can reduce the availability of nutrients to organisms, making the water body even less productive than an oligotrophic one.

Table 1.1 Some general characteristics of oligotrophic and eutrophic lakes.

Characteristic	Oligotrophic	Eutrophic
primary production	low	high
diversity of primary producers	high species diversity, with low population densities	low species diversity, with high population densities
light penetration into water column	high	low
toxic blooms	rare	frequent
plant nutrient availability	low	high
animal production	low	high
oxygen status of surface water	high	low
fish	salmonid fish (e.g. trout, char) often dominant	coarse fish (e.g. perch, roach, carp) often dominant

❍ Why is light penetration poor in eutrophic lakes?

● The high density of phytoplankton absorbs light for photosynthesis and prevents it penetrating deeper into the water.

Table 1.2 Trophic bands for standing waters. (Phosphorus concentrations tend to be higher in running waters that carry suspended sediment.)

Trophic band	Total phosphorus/mg l^{-1}
dystrophic	<0.005
oligotrophic	0.005–0.01
mesotrophic	0.01–0.03
eutrophic	0.03–0.1
hypertrophic	>0.1

More recently, **trophic bands** have been defined in relation to levels of nutrients measured by chemical analysis. Table 1.2 shows trophic bands as defined in relation to concentrations of total phosphorus.

The trophic state of water bodies and rivers varies depending on a number of factors, including position in the landscape and management of surrounding land. In general, upland areas are more likely to have nutrient-poor (oligotrophic) water, characterized by relatively fast-flowing rivers (Figure 1.1) and lakes that have clear water with limited higher plant communities. By contrast, lowland waters in more fertile river catchments tend to be nutrient-rich (eutrophic), and lakes in lowland areas are more likely to be turbid with lush fringing vegetation. Lowland rivers have slower flow and are likely to be more nutrient rich as a result of soluble compounds having been washed into them. They are likely to have fringing vegetation and some floating and submerged aquatic plants (Figures 1.2 and 1.3). In aquatic systems, the term macrophyte is used to describe any large plant (*macro*, large; *phyte,* plant). The term is used to distinguish angiosperms (whether emergent, floating or submerged) from small algae such as **diatoms** (which are strictly not plants at all, but are often lumped together with plants when considering the productivity of ecosystems).

❍ What is the process by which nutrient elements are lost from the soil profile by the action of excess rainfall draining through it, which may eventually deliver them to a surface water body?

● Leaching.

The term 'eutrophication' came into common usage from the 1940s onwards, when it was realized that, over a period of years, plant nutrients derived from industrial activity and agriculture had caused changes in water quality and the biological character of water bodies. In England and Wales, eutrophication has been a particular concern since the late 1980s, when public awareness of the problem was heightened by widespread toxic blue–green bacterial blooms (commonly, but incorrectly, referred to as algal blooms) in standing and slow-flowing freshwaters. Figure 1.4 shows blue–green bacteria (cyanobacteria) growing at the margins of a lake. Cyanobacteria are not typical bacteria, not only because some of them are photosynthetic, but also because some of them can be multicellular, forming long chains of cells. Nonetheless, cyanobacteria clearly belong to the kingdom Bacteria because of their internal cellular structure.

- ❍ Why are cyanobacteria so productive in eutrophic water bodies (Figure 1.4) compared with oligotrophic ones?
- ● The ready availability of nutrients allows rapid growth. In oligotrophic water the rate of growth is limited by the nutrient supply, but in eutrophic water it is often only the availability of light which regulates primary production.

Figure 1.1 Fast-flowing upland stream with clear water and few plants.

Figure 1.2 Lowland river, rich in aquatic plant species.

Figure 1.3 Rich community of macrophytes. The tall plants growing out of the water are described as emergent.

Figure 1.4 A cyanobacterial bloom.

1.3 Resource availability and species diversity

A wide range of ecosystems has been studied in terms of their species diversity and the availability of resources. Each produces an individual relationship between these two variables, but a common pattern emerges from most of them, especially when plant diversity is being considered. This pattern has been named the humped-back relationship and suggests diversity is greatest at intermediate levels of productivity in many systems (Figure 1.5).

- ❍ Recall how species diversity differs from species richness.
- ● Species diversity includes a measure of how evenly spread the biomass is between species (equitability) rather than a simple count of the species present.

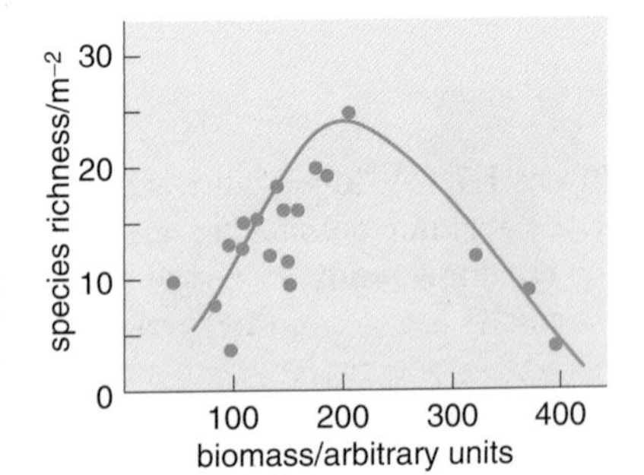

Figure 1.5 The species richness of samples of vegetation from South Africa shows a classic humped-back relationship with ecosystem productivity as inferred from amount of biomass per unit area.

An explanation for this relationship is that at very low resource availability, and hence ecosystem productivity, only a limited number of species are suitably adapted to survive. As the limiting resource becomes more readily available, then more species are able to grow. However, once resources are readily available, then the more competitive species within a community are able to dominate it and exclude less vigorous species.

In most ecosystems it is the availability of mineral nutrients (especially nitrogen and phosphorus) that limits productivity. In eutrophic environments these nutrients are readily available by definition, so species diversity can be expected to be lower than in a more mesotrophic situation. It is for this reason that eutrophication is regarded as a threat to biodiversity. Eutrophication of the environment by human-mediated processes can have far reaching effects, because the nutrients released are often quite mobile. Together with habitat destruction, it probably represents one of the greatest threats to the sustainability of biodiversity over most of the Earth.

1.4 Natural eutrophication

Eutrophication of habitat can occur without human interference. Nutrient enrichment may affect habitats of any initial trophic state, causing distinctive changes to plant and animal communities. The process of primary succession is normally associated with a gradual eutrophication of a site as nutrients are acquired and stored by vegetation both as living tissue and organic matter in the soil.

There is a long-standing theory that most water bodies go through a gradual process of nutrient enrichment as they age: a process referred to as **natural eutrophication**. All lakes, ponds and reservoirs have a limited lifespan, varying from a few years for shallow water bodies to millions of years for deep crater lakes created by movements of the Earth's crust. They fill in gradually with sediment and eventually became shallow enough for plants rooted in the bed sediment to dominate, at which point they develop into a closed swamp or fen and are eventually colonized by terrestrial vegetation (Figures 1.6 and 1.7).

Figure 1.6 Cross-section through a water body that is gradually becoming filled with silt deposits and organic matter as a result of vegetation growth.

Figure 1.7 A floodplain water body becoming colonized with emergent macrophytes, which may eventually cause it to disappear through an accumulation of silt and organic matter.

Nutrient enrichment occurs through addition of sediment, rainfall and the decay of resident animals and plants and their excreta. Starting from an oligotrophic state with low productivity, a typical temperate lake increases in productivity fairly quickly as nutrients accumulate, before reaching a steady state of eutrophy which might last for a very long time (perhaps thousands of years). However, it is possible for the nutrient status of a water body to fluctuate over time and for trophic state to alter accordingly. Study of sediments in an ancient lake in Japan, Lake Biwa (believed to be around four million years old) suggests that it has passed through two oligotrophic phases in the last half million years, interspersed with two mesotrophic phases and one eutrophic phase. Evidence such as this has led to the suggestion that the nutrient status of lakes reflects contemporary nutrient supply, and can increase or decrease in response to this. The processes by which nutrients are washed downstream or locked away in sediments help to ensure that reversal of natural eutrophication can occur.

Rivers vary in trophic state between source and sea, and generally become increasingly eutrophic as they approach sea-level.

1.5 Human-induced eutrophication

While eutrophication does occur independently of human activity, increasingly it is caused, or amplified, by human inputs. Human activities are causing pollution of water bodies and soils to occur to an unprecedented degree, resulting in an array of symptomatic changes in water quality and in species and communities of associated organisms. In 1848 W. Gardiner produced a flora of Forfarshire, in which he described the plants growing in Balgavies Loch. He talked of 'potamogetons [pondweeds] flourishing at a great depth amid the transparent waters, animated by numerous members of the insect and finny races'. These 'present a delightful spectacle, and the long stems of the white and yellow water lilies may be traced from their floating flowers to the root'. By 1980, the same loch had very low transparency and dense growths of planktonic algae throughout the summer. The submerged plants grew no deeper than 2 m, and in the 1970s included just three species of *Potamogeton*,where previously there were 17.

For any ecosystem, whether aquatic or terrestrial, nutrient status plays a major part in determining the range of organisms likely to occur. Characteristic assemblages of plant and associated animal species are found in water with different trophic states. Table 1.3 shows some of the aquatic macrophyte species associated with different concentrations of phosphorus in Britain.

Table 1.3 Concentrations of phosphorus (in rivers) with which plant species are correlated.

Phosphorus present as soluble reactive phosphorus (SRP)*/mg P l^{-1}	Plant species (see Figure 1.8 for illustrations)
<0.1	bog pondweed, *Potamogeton polygonifolius* river water-crowfoot, *Ranunculus fluitans*
0.1–0.4	fennel-leaved pondweed, *Potamogeton pectinatus*
0.4–1.0	yellow water-lily, *Nuphar lutea* arrowhead, *Sagittaria sagittifolia*
>1.0	spiked water-milfoil, *Myriophyllum spicatum*

* This term is explained in Section 2.1.

- ❍ What impression would you gain from an observation that a population of river water-crowfoot in a particular stretch of river had been largely replaced by fennel-leaved pondweed over a three-year period?
- ● The phosphorus concentration of the water may have increased.

(a)

(b)

(c)

Figure 1.8 Three aquatic macrophyte species which differ in their tolerance to eutrophication: (a) river water-crowfoot (*Ranunculus fluitans*) is intolerant, (b) yellow water-lily (*Nuphar lutea*) is intermediate and (c) spiked water-milfoil (*Myriophyllum spicatum*) is tolerant.

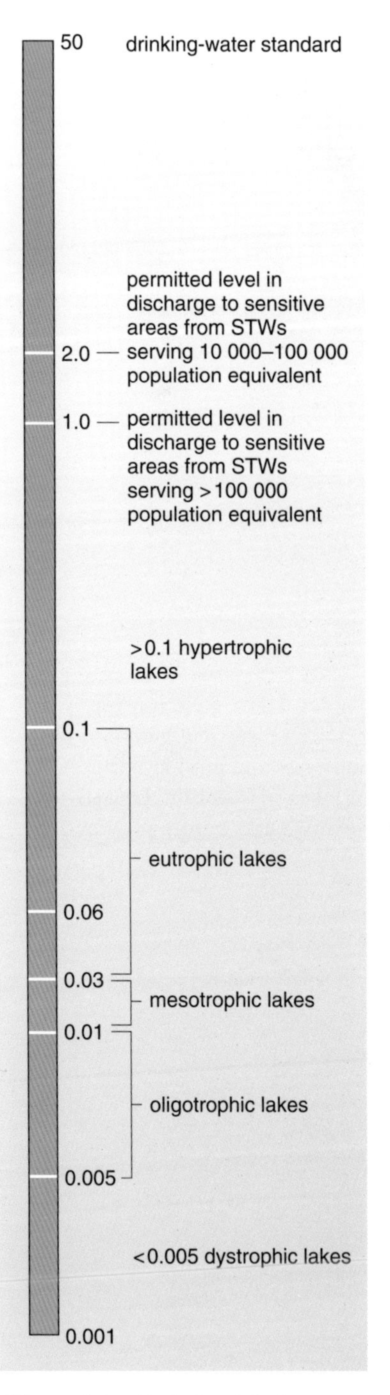

Figure 1.9 Relationship between levels (in mg l^{-1}) of total phosphorus in standing water and the nutrient status of lakes (STW, sewage treatment works).

Figure 1.9 illustrates the relationship between levels of total phosphorus in standing water and the nutrient status of lakes. Above a level of 0.1 mg phosphorus per litre, biodiversity often declines. Using the trophic bands defined in Table 1.2, this is the concentration at which lakes are considered to become hypertrophic. This is way below the standard of 50 mg l^{-1} set as the acceptable limit for phosphorus in drinking water. Nutrient loadings this high are generally caused by human activities. Extremely high levels of eutrophication are often associated with other forms of pollution, such as the release of toxic heavy metals, resulting in ecosystems that may no longer support life (Figure 1.10).

Figure 1.10 Polluted river in an industrial area.

For lakes with no written historical records, the diatom record of sediments can be used to study earlier periods of natural change in water quality, and to provide a baseline against which to evaluate trends in artificial or human-induced eutrophication. Diatoms are microscopic photosynthetic organisms (algae of the kingdom Protoctista), which live either free-floating in lakes or attached to the surface of rocks and aquatic vegetation. It is well established that some species of diatom can tolerate oligotrophic conditions whereas others flourish only in more eutrophic waters. When they die, their tiny (< 1 mm) bony capsules, which can be identified to species level, sink to the bed and may be preserved for thousands of years. A historical record of which species have lived within a water body can therefore be constructed from an analysis of a core sample taken from its underlying sediment.

Studies of diatom remains have demonstrated that current levels of eutrophication far exceed those found historically. In the English Lake District, productivity and sediment input increased in some lakes when vegetation was cleared by Neolithic humans around 5000 years ago, and again when widespread deforestation occurred 2000 years ago. However the greatest increases in productivity, sediment levels and levels of carbon, nitrogen and phosphorus, have occurred since 1930. Figure 1.11 shows the general pattern of changes in productivity in Cumbrian lakes through history as the type and intensity of human activities has changed.

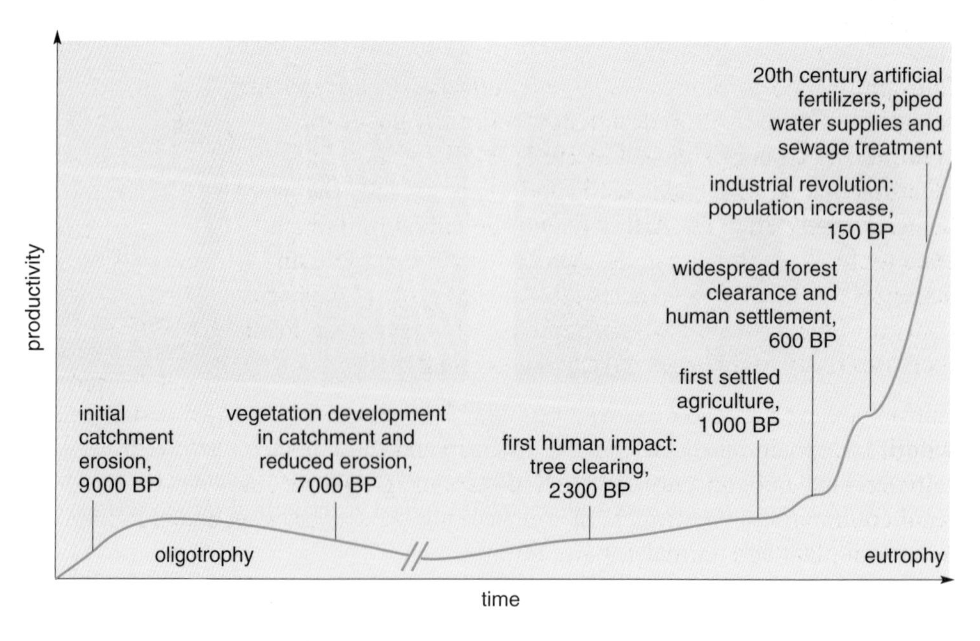

Figure 1.11 Relationship between historical human activities and productivity of lakes in Cumbria, UK. (BP means 'years before present'.)

In the Norfolk Broads, the waters of the River Ant had a diverse macrophyte flora during the 19th century. The submerged species known as water soldier (*Stratiotes aloides*, Figure 1.12) was common, but by 1968 the only macrophytes remaining were those with permanently floating leaves, such as water-lilies. During that period, throughout the Broads, there was a general trend away from clear-water habitats, typified by, for example, the diminutive angiosperm known as the holly-leaved naiad (*Najas marina*), towards habitats containing more productive species, such as pondweeds (*Potamogeton* spp.) and hornworts (*Ceratophyllum* spp.). In some cases, they eventually became eutrophic habitats with turbid water, typified by free-floating green algae and cyanobacteria, with very few macrophytes at all. For example, hornwort (*Ceratophyllum demersum*) was almost choking Alderfen Broad in 1963, but had almost disappeared by 1968 to be replaced eventually by algal blooms in the 1990s.

Sediment cores from the River Ant and neighbouring broads suggest that observed changes in plant community composition were linked to rising levels of total phosphorus: mean levels in the area rose dramatically between 1900 and 1975 (Figure 1.13), but have since fallen as a result of actions taken to remove phosphorus from the system.

Figure 1.12 A free-floating macrophyte, water soldier (*Stratiotes aloides*), spends most of its life submerged, but rises to the surface to flower.

- ❍ Using the trophic bands in Table 1.2, describe the change in the River Ant broads between 1800 and 1975.
- ● In 1800 the water was at the upper end of the oligotrophic range; it had moved through the mesotrophic range to become eutrophic by 1900, and by 1940 would be classed as hypertrophic. Between 1940 and 1975 there was a further threefold increase in the concentration of total phosphorus.

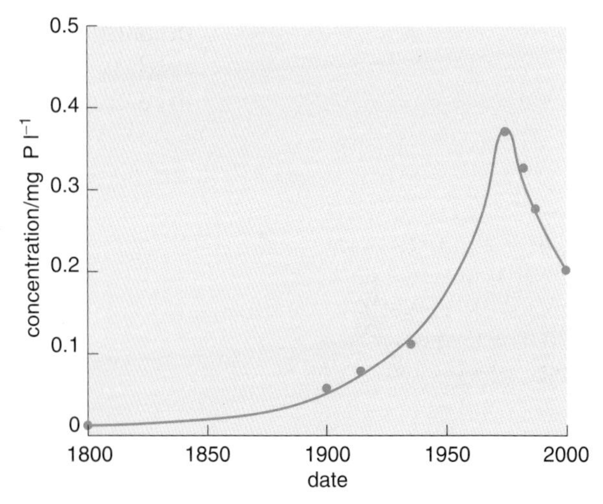

Figure 1.13 Annual peak concentration of total phosphorus in the water bodies linked to the River Ant, Norfolk Broads, England, where a programme of remedial action to address eutrophication has been implemented over the past 25 years.

Eutrophication has damaged a large number of sites of special scientific interest (SSSIs) designated in the UK under the Wildlife and Countryside Act of 1981: English Nature has identified a total of 90 lake SSSIs and 12 river SSSIs that have been adversely affected. Artificial eutrophication in rivers is even more widespread than in lakes and reservoirs. Human activities worldwide have caused the nitrogen and phosphorus content of many rivers to double and, in some countries, local increases of up to 50 times have been recorded.

Eutrophication has also become a problem for terrestrial wildlife. Deposition of atmospheric nitrogen and the use of nitrogen-rich and phosphorus-rich fertilizers in agriculture has resulted in nutrient enrichment of soils and has caused associated alteration of terrestrial plant and animal communities.

Some of the effects of large-scale eutrophication have adverse consequences for people, and efforts to manage or reduce eutrophication in different countries now cost substantial sums of money. Removing nitrates from water supplies in England and Wales cost £20 million in 1995. Higher frequency of algal blooms increases the costs of filtration for domestic water supply and may cause detectable tastes and odours due to the secretion of organic compounds. If the bloom is large, these compounds can accumulate to concentrations that are toxic to mammals and sometimes fish. Furthermore, the high productivity of the blooms means that although oxygen is released by photosynthesis during the day, the effect of billions of cells respiring overnight can deplete the water of oxygen, resulting in fish dying through suffocation even if they tolerate the toxins.

❍ Are fish most at risk from suffocation in warm or cool water? (You may wish to refer back to Block 3.)

● Warm water, because oxygen is less soluble at warmer temperatures and is therefore more rapidly depleted by respiring organisms, especially as respiration rate also increases with temperature.

Another problem caused to the water industry by algal blooms is the production of large quantities of fine organic detritus, which, when collected within waterworks' filters, may support clogging communities of aquatic organisms such as nematode worms, sponges and various insects. These may subsequently find their way into water distribution pipes and on occasion appear in tap water!

Question 1.1

Referring back to the concept of Grime's life strategies in Block 3, Part 2 *Life*, consider the differences in nutrient acquisition and use between his three principal categories: competitors, stress tolerators and ruderals. In terrestrial systems, which life strategy is likely to be most successful in (a) oligotrophic and (b) eutrophic conditions?

Effects of eutrophication

2

A number of biological changes may occur as a result of eutrophication. Some of these are direct (e.g. stimulation of algal growth in water bodies), while others are indirect (e.g. changes in fish community composition due to reduced oxygen concentrations). This section summarizes some of the typical changes observed in aquatic, marine and terrestrial ecosystems following eutrophication.

Some typical changes observed in lakes following artificial eutrophication are summarized in Table 2.1. Similar characteristic changes are observed in other freshwater systems.

Table 2.1 Typical changes observed in lakes experiencing 'artificial' eutrophication.

- Turbidity increases, reducing the amount of light reaching submerged plants.
- Rate of sedimentation increases, shortening the lifespan of open water bodies such as lakes.
- Primary productivity usually becomes much higher than in unpolluted water and may be manifest as extensive algal or bacterial blooms.
- Dissolved oxygen in water decreases, as organisms decomposing the increased biomass consume oxygen.
- Diversity of primary producers tends to decrease and the dominant species change. Initially the number of species of green algae increases, causing temporary increase in diversity of primary producers. However, as eutrophication proceeds, blue–green bacteria become dominant, displacing many algal species. Similarly some macrophytes (e.g. bulrushes) respond well initially, but due to increased turbidity and anoxia (reduced oxygen) they decline in diversity as eutrophication proceeds.
- Fish populations are adversely affected by reduced oxygen availability, and the fish community becomes dominated by surface-dwelling coarse fish, such as pike (*Esox lucius,* see Figure 2.1) and perch (*Perca fluviatilis*).
- Zooplankton (e.g. *Daphnia* spp.), which eat phytoplankton, are disadvantaged due to the loss of submerged macrophytes, which provide their cover, thereby exposing them to predation.
- Increased abundance of competitive macrophytes (e.g. bulrushes) may impede water flow, increasing rates of silt deposition.
- Drinking water quality may decline. Water may be difficult to treat for human consumption, for example due to blockage of filtering systems. Water may have unacceptable taste or odour due to the secretion of organic compounds by microbes.
- Water may cause human health problems, due to toxins secreted by the abundant microbes, causing symptoms that range from skin irritations to pneumonia.

Figure 2.1 The pike (*Esox lucius*), a species of fish that can tolerate eutrophic water.

In oligotrophic systems, even quite small increases in nutrient load can have relatively large impacts on plant and animal communities.

2.1 Effects on primary producers in freshwater ecosystems

Plant species differ in their ability to compete as nutrient availability increases. Some floating and submerged macrophyte species are restricted to nutrient-poor waters, while others are typical of nutrient-rich sites (see Table 2.2). Figure 2.2 shows turbid water in a polluted drainage ditch associated with localized growth of algae. There are no aquatic plants present.

Table 2.2 Species restricted to nutrient-poor or typical of nutrient-rich standing water.

Trophic state	Associated macrophyte species
oligotrophic	alternate water-milfoil (*Myriophyllum alternifolium*) bog pondweed (*Potamogeton polygonifolius*)
oligo–mesotrophic	bladderwort (*Utricularia vulgaris*)
eutrophic	hairlike pondweed (*Potamogeton trichoides*)
tending towards hypertrophic	spiked water-milfoil (*Myriophyllum spicatum*) fennel-leaved pondweed (*Potamogeton pectinatus*)

Figure 2.2 A polluted drainage ditch with luxuriant growth of algae, forming a carpet over its surface.

In rivers, the presence of plant species such as the yellow water-lily (*Nuphar lutea*) and the arrowhead (*Sagittaria sagittifolia,* Figure 2.3) are likely to indicate eutrophic conditions. In some rivers, the fennel-leaved pondweed (*Potamogeton pectinatus*) is tolerant of both sewage and industrial pollution.

Whereas some species can occur in waters with quite a wide range of nutrient levels, some are relatively obligate to specific trophic bands and are unable to survive if nutrient levels alter significantly from those to which they are adapted. In 1989, Michael Jeffries derived ranges of tolerance for a number of macrophyte species by studying literature on their occurrence and distribution in relation to different aspects of water quality. He also reviewed results of scientific studies reported in the literature to determine what concentrations of nitrate, ammonia, phosphorus, suspended solids and biological oxygen demand (BOD) appeared to be associated with severe or total loss of macrophyte species due to eutrophication (see Table 2.3). Research has suggested that changes to certain macrophyte communities can occur at soluble reactive phosphorus concentrations as low as 20 μg l^{-1} (0.02 mg l^{-1}). Soluble reactive phosphorus (SRP) is the term commonly used to describe phosphorus that is readily available for uptake by organisms. It is used in contrast to measures of total phosphorus, which include forms of the element that are bound to sediment particles or locked up in large organic molecules. These forms are unavailable for immediate uptake, but they may become available over time.

Figure 2.3 Arrowhead (*Sagittaria sagittifolia*), a macrophyte species that is tolerant of eutrophic conditions.

Question 2.1

Water samples from two lowland rivers, A and B, are found to contain the following concentrations of plant nutrients.

Nutrient	Concentration/mg l^{-1} River A	River B
nitrate	2.2	12.1
ammonia	0.07	0.6
SRP	0.18	0.13

By reference to Table 2.3, what conclusions can you draw about the probable diversity of aquatic macrophytes in each of the rivers?

Table 2.3 Typical concentrations (in mg l^{-1}) of selected water quality parameters associated with 'natural', degraded and severe loss of plant species.

Condition	SRP/ mg P l^{-1}	Nitrate/ mg N l^{-1}	Ammonia/ mg N l^{-1}	Suspended solids	BOD
'natural'	<0.1	<3.0	<0.2	<30	<2.0
degraded (partial loss of species found under 'natural' conditions)	0.1–0.2	3.0–10	0.2–5.0	30–100	2.0–6.0
severe loss of species	>0.2	>10	>5.0	>100	>6.0

BOD, biological oxygen demand; SRP, soluble reactive phosphorus.

2.1.1 Loss of submerged plant communities

One of the symptoms of extreme eutrophication in shallow waters is often a substantial or complete loss of submerged plant communities and their replacement by dense phytoplankton communities (algal blooms). This results not only in the loss of characteristic plant species (macrophytes) but also in reduced habitat structure within the water body. Submerged plants provide refuges for invertebrate species against predation by fish. Some of these invertebrate species are phytoplankton-grazers and play an important part in balancing relative proportions of macrophytes and phytoplankton. Submerged macrophytes also stabilize sediments and the banks of slow-flowing rivers or lakes. Bodies of water used for recreation (boating for example) become more vulnerable to bank destabilization and erosion in the absence of well-developed plant communities, making artificial bank stabilization necessary (Figure 2.4). Submerged plants also have a role in the oxygenation of lower water layers and in the maintenance of aquatic pH.

Figure 2.4 River bank stabilization.

❍ Name three species of submerged macrophytes that are tolerant of eutrophic water.

● Spiked water-milfoil (*Myriophyllum spicatum*), fennel-leaved pondweed (*Potamogeton pectinatus*) and arrowhead (*Sagittaria sagittifolia*).

2.1.2 Algal blooms

The enrichment of water bodies by eutrophication may be followed by population explosions or 'blooms' of planktonic organisms.

❍ Bursts of primary production in an aquatic ecosystem in response to an increased nutrient supply are commonly referred to as 'algal blooms'. Explain why this term is taxonomically incorrect.

● The organisms responsible may be either algae or bacteria, or a mixture of the two. It is incorrect to refer to bacteria as algae as they belong to a completely different taxonomic kingdom.

'Algal blooms' are a well-publicized problem associated with increased nutrient levels in surface waters. The higher the concentration of nutrients, the greater the primary production that can be supported. Opportunistic species like some algae are able to respond quickly, showing rapid increases in biomass. Decomposition of these algae by aerobic bacteria depletes oxygen levels, often very quickly. This can deprive fish and other aquatic organisms of their oxygen supply and cause high levels of mortality, resulting in systems with low diversity. The odours associated with algal decay taint the water and may make drinking water unpalatable. Species of cyanobacteria that flourish in nutrient-rich waters can produce powerful toxins that are a health hazard to animals. Such problems are well documented for a number of famous lakes. The Zurichsee in Switzerland has been subject to seasonal blooms of the cyanobacterium *Oscillatoria rubescens* due to increased sewage discharge from new building developments on its shores. For lakes in Wisconsin, USA, 'nuisance' blooms of algae or bacteria occur whenever concentrations of phosphate and nitrate rise.

2.2 Effects on consumers in freshwater ecosystems

Increased productivity tends to increase rates of deoxygenation in the surface layer of lakes. Although phytoplankton release oxygen to the water as a by-product of photosynthesis during the day, water has a limited ability to store oxygen and much of it bubbles off as oxygen gas. At night, the phytoplankton themselves, the zooplankton and the decomposer organisms living on dead organic matter are all respiring and consuming oxygen. The store of dissolved oxygen thus becomes depleted and diffusion of atmospheric oxygen into the water is very slow if the water is not moving.

❍ Refer back to Block 3 for the relative rate of oxygen diffusion in water compared with its rate in air.

● Oxygen diffuses through water at approximately one ten-thousandth of its rate through air.

Still waters with high productivity are therefore likely to become anoxic.

Figures 2.5 and 2.6 give an example of the change in aquatic invertebrate species following eutrophication. In unpolluted water, mayfly larvae may be found. In polluted water, these species cannot survive due to reduced oxygen availability and are likely to be replaced by species, such as the bloodworm, which can tolerate lower oxygen concentrations.

(a) (b)

Figure 2.5 (a) Mayfly larva, found in unpolluted water. (b) Adult mayfly.

Figure 2.6 Pollution-tolerant species, the bloodworm.

Figure 2.7 Roach, a species of coarse fish (a cyprinid) able to benefit from eutrophic conditions.

Many species of **coarse fish**, such as roach (*Rutilus rutilus*, a **cyprinid** fish, Figure 2.7), can also tolerate low oxygen concentrations in the water, sometimes gulping air, and yields of fish may indeed increase due to the high net primary production (NPP) of the system. However these species are generally less desirable for commercial fishing than others such as salmon (*Salmo salar*, a **salmonid** fish, Figure 2.8), which depend on cool, well-oxygenated surface water. Populations of such species usually decline in waters that become eutrophic (Figure 2.9); they may be unable to live in a deoxygenated lake at all, resulting in fish kills (Figure 2.10). They may also be unable to migrate through deoxygenated waters to reach spawning grounds, resulting in longer-term population depressions.

Figure 2.8 Salmon (*Salmo salar*), a salmonid fish intolerant of eutrophication.

Lake Victoria is one of the world's largest lakes and used to support diverse communities of species **endemic** to the lake (i.e. species that are found only there), but it now suffers from frequent fish kills caused by episodes of deoxygenation. In the 1960s deoxygenation was limited to certain areas of the lake, but it is now widespread. It is usually associated with at least a tenfold increase in the algal biomass and a fivefold increase in primary productivity.

When eutrophication reaches a stage where dense algal growth outcompetes marginal aquatic plants, even relatively tolerant fish species suffer from the consequent loss of vegetation structure, especially young fish (Figure 2.11). Spawning is reduced for fish species that attach their eggs to aquatic plants or their detritus, and fish that feed on large plant-eating invertebrates, such as snails and insect nymphs, suffer a reduced food supply.

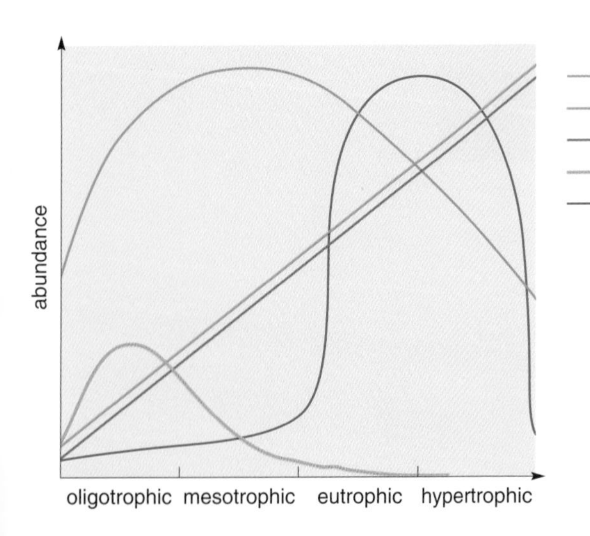

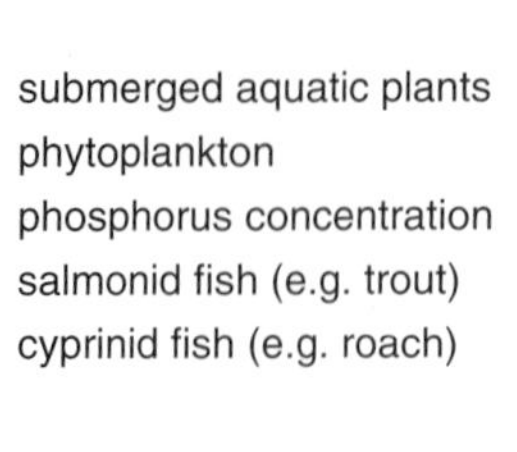

Figure 2.9 Changes occurring in northern temperate lakes due to eutrophication.

Figure 2.10 Fish kill caused by deoxygenation of the water.

○ In a southeast Asian village where cyprinid fish from the local pond are an important source of protein, eutrophication of the water by domestic sewage is seen as advantageous. Why?

● The cyprinid fish are tolerant of deoxygenation, and the increased NPP boosts their food supply; therefore the yield of fish improves.

Figure 2.11 Newly-hatched salmon are particularly demanding of oxygen-rich water.

2.3 Effects on terrestrial vegetation

○ Recall from Topic 2 *Atmospheric Chemistry and Pollution* why nitrogen is becoming increasingly available to terrestrial ecosystems in many parts of the world.

● Emission of nitrogen oxides from burning fossil fuel and of ammonia from intensive agriculture result in nitrogen compounds being transported and deposited by atmospheric processes.

Some suggested global scenarios for the year 2100 identify nitrogen deposition (together with land use and climate change) as one of the most significant 'drivers' of biodiversity change in terrestrial ecosystems.

Atmospheric deposition of nitrogen, together with the deposition of phosphorus-rich sediments by floods, can alter competitive relationships between plant species within a terrestrial community. This can cause significant changes in community composition, as species differ in their relative responses to elevated nutrient levels. As is the case with aquatic vegetation, terrestrial species that are able to respond to extra nitrogen and phosphorus with elevated rates of photosynthesis will achieve higher rates of biomass production, and are likely to become increasingly dominant in the vegetation. Atmospheric deposition of nutrients can reduce, or even eliminate, populations of species that have become adapted to low nutrient conditions and are unable to respond to increased nutrient availability. Some vegetation communities of conservation interest are directly threatened by atmospheric pollution.

In Britain, rare bryophytes are found associated with snowbeds (Figure 2.12). Most of the late-lying snowbeds in Britain are in the Central Highlands of Scotland, which are also areas of very high deposition of nitrogenous air pollutants. Snow is a very efficient scavenger of atmospheric pollution and melting snowbeds release their pollution load at high concentrations in episodes known as 'acid flushes'. The flush of nitrogen is received by the underlying vegetation when it has been exposed following snowmelt. Concentrations of nutrients in the meltwater of Scottish snowbeds have already been shown to damage underlying bryophytes, including a rare species called *Kiaeria starkei*. Recovery from damage is slow, and sometimes plants show no signs of recovery even four weeks after exposure to polluted meltwater. Given the very short growing season, this persistent damage can greatly reduce the viability and survival of the plants. Tissue nitrogen concentrations in *Kiaeria starkei* have

Figure 2.12 Typical habitat for *Kiaeria starkei* is a late-lying snowbed, such as this in the Scottish Cairngorms, where the meltwater in spring will carry a flush of atmospheric pollutants, including nitrogen compounds.

been shown to be up to 50% greater than that recorded in other upland bryophytes. This example emphasizes the potential threat of atmospheric pollution to snowbed species, and suggests that some mountain plant communities may receive much higher pollution loadings than was previously realized.

The deposition of atmospheric nitrogen can be enhanced at high altitude sites as a consequence of cloud droplet deposition on hills. Sampling of upland plant species at sites in northern Britain has shown marked increases in nitrogen concentration in leaves with increasing nitrogen deposition, which is, in turn, correlated with increasing altitude. The productivity of the species was also found to increase in line with the amount of nitrogen deposited. Plant species can therefore respond directly to elevated levels of nitrogen. In the longer term, the relative dominance of species is likely to alter depending on their ability to convert elevated levels of deposited nitrogen into biomass.

❍ What will be the effect on species diversity of increasing biomass?

● As biomass increases beyond an optimal value, species diversity will decline.

Atmospheric pollution can also affect plant–insect interactions. Unusual episodes of damage to heather moorland in Scotland have been caused by the winter moth (*Operophtera brumata*) in recent years. It has been suggested that this may be due to the effects of increased nitrogen supply on heather plants, including increased shoot growth and a decrease in the carbon : nitrogen ratio in plant tissues. Winter moth larvae have been shown to grow faster on nitrogen-treated heather plants, so it is possible that increased atmospheric deposition of nitrogen may have a role in winter moth outbreaks and the associated degradation of heather moorland in upland Britain.

Although uplands are more susceptible to atmospheric deposition of nitrogen, the effects can be seen in lowland areas too. Nitrogen deposition and the consequent eutrophication of ecosystems is now regarded as one of the most important causes of decline in plant species in the Netherlands. Figure 2.13 shows how the number of grassland species of conservation interest in south Holland declines as the nitrogen load increases. The maximum percentage of species (approximately 95%) is possible at a nitrogen load of about 6 kg N ha^{-1} yr^{-1}. At loads higher than 10 kg N ha^{-1} yr^{-1} the number of species declines due to eutrophication effects, and below 5 kg ha^{-1} yr^{-1} nitrogen may be too limiting for a few species.

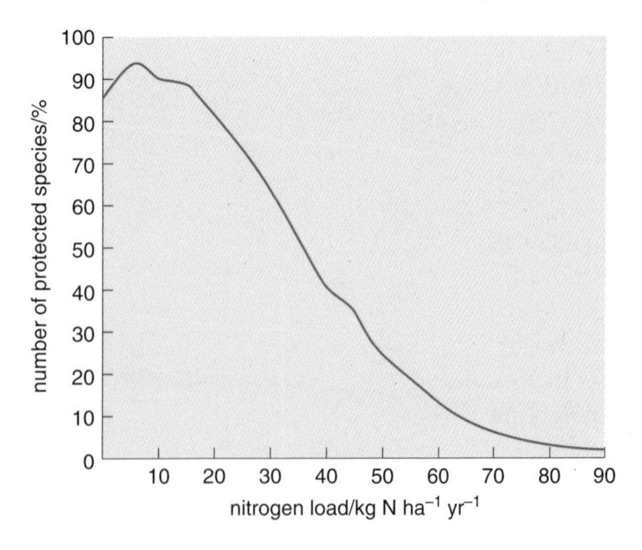

Figure 2.13 Relationship between potential number of protected grassland species in grassland and nitrogen load in South Holland.

A significant proportion of important nature conservation sites in Britain are subject to nitrogen and/or sulfur deposition rates that may disturb their biological communities. Lowland heath ecosystems, for example, have a high profile for conservation action in Britain. They typically have low soil nutrient levels and a vegetation characterized by heather (*Calluna vulgaris*). Under elevated atmospheric deposition of nitrogen, they tend to be invaded by taller species, including birch (*Betula* spp., Figure 2.14), bracken (*Pteridium aquilinum*) and the exotic invader, rhododendron (*Rhododendron ponticum*).

Figure 2.14 Invasion of lowland heath by birch trees.

Figure 2.15 Otter, *Lutra lutra*.

- ❍ Recall from Block 3, Part 2 *Life*, Grime's concept of life strategy. What are the strategies of (a) heather and (b) birch?
- ● (a) Heather is a stress tolerator, adapted to survive in a difficult environment, where nutrients are very scarce. (b) Birch is a more competitive species, able to utilize additional nutrients when available, to increase its growth rate.

A large number of SSSIs in the UK, designated as such on account of their terrestrial plant communities, are considered to have been damaged by eutrophication. This has been identified as a factor in the decline of some important UK habitats, including some identified for priority action under the UK's Biodiversity Action Plan (BAP). Wet woodlands, for example, occur on poorly drained soils, usually with alder, birch and willow as the predominant tree species, but sometimes oak, ash or pine occur in slightly drier locations. These woodlands are found on floodplains, usually as a successional habitat on fens, mires and bogs, along streams and in peaty hollows. They provide an important habitat for a variety of species, including the otter (*Lutra lutra,* Figure 2.15), some very rare beetles and craneflies. They also provide damp microclimates, which are particularly suitable for bryophytes, and have some unusual habitat features not commonly found elsewhere, such as log jams in streams which support a rare fly, *Lipsothrix nigristigma*. Wet woodlands occur on a range of soil types, including relatively nutrient-rich mineral soils as well as acid, nutrient-poor ones. Nevertheless, many have been adversely affected by eutrophication, resulting in altered ground flora composition and changes in the composition of invertebrate communities.

Nutrient enrichment can also affect habitats found in drier sites. Eutrophication caused by runoff from adjacent agricultural land has been identified as a cause of altered ground flora composition in upland mixed ash woods for example. These woods are notable for bright displays of flowers such as bluebell (*Hyacinthoides non-scripta*), primrose (*Primula vulgaris*) and wild garlic (*Allium ursinum*, Figure 2.16a). They also support some very rare woodland flowers which are largely restricted to upland ash woods, such as dark red helleborine (*Epipactis atrorubens*, Figure 2.16b) and Jacob's ladder (*Polemonium caeruleum*).

(a)

(b)

Figure 2.16 (a) Wild garlic (*Allium ursinum*) and (b) dark-red helleborine (*Epipactis atrorubens*) are species of the woodland floor that may be displaced if nutrient availability in the soil increases.

Other terrestrial UK BAP habitats that may be adversely affected by nutrient enrichment from agricultural fertilizers or atmospheric deposition include lowland wood pasture, lowland calcareous grassland, upland hay meadows and lowland meadows; again the result can be altered plant species composition.

Coastal marshes and wetlands in many parts of the world have been affected by invasion of 'weed' or 'alien' species. Eutrophication can accelerate invasion of aggressive, competitive species at the expense of slower growing native species. In the USA, many coastal marshes have been invaded by the common reed (*Phragmites australis*, Figure 2.17). *Phragmites* is a fierce competitor and can outcompete and entirely displace native marsh plant communities, causing local extinction of plants and the insects and birds that feed on them. *Phragmites* can spread by underground rhizomes and can rapidly colonize large areas. However, it is the target of conservation effort in some areas, including Britain, because the reedbeds it produces provide an ideal habitat for rare bird species such as the bittern (*Botaurus stellarus*). But its spread is not always beneficial for nature conservation, as it often results in the drying of marsh soils, making them less suitable for typical wetland species and more suitable for terrestrial species. This is because *Phragmites* is very productive and can cause ground levels to rise due to deposition of litter and the entrapment of sediment. Thus eutrophication can also play an indirect part in the loss of wetland habitats.

Figure 2.17 An extensive stand of the common reed (*Phragmites australis*), which is found around the globe in nutrient-rich, shallow freshwater. It is expanding its extent in many areas in response to eutrophication of previously nutrient-poor ecosystems.

2.4 Effects on marine systems

In the marine environment, nutrient enrichment is suspected when surface phytoplankton blooms are seen to occur more frequently and for longer periods. Some species of phytoplankton release toxic compounds and can cause mass mortality of other marine life in the vicinity of the bloom. Changes in the relative abundance of phytoplankton species may also occur, with knock-on effects throughout the food web, as many zooplankton grazers have distinct feeding preferences. In sheltered estuarine areas, high nutrient levels appear to favour the growth of green macroalgae ('seaweeds') belonging to such genera as *Enteromorpha* and *Ulva* (Figure 2.18).

Figure 2.18 The macroalga *Ulva taeniata,* which can grow to several metres in length, given a sufficient supply of nutrients.

2.4.1 Estuarine species

Nutrient runoff from the land is a major source of nutrients in estuarine habitats. Shallow-water estuaries are some of the most nutrient-rich ecosystems on Earth, due to coastal development and the effects of urbanization on nutrient runoff. Figure 2.19 shows some typical nitrogen pathways. Nitrogen loadings in rainfall are typically assimilated by plants or denitrified, but septic tanks tend to add nitrogen below the reach of plant roots, and if situated near the coast or rivers can lead to high concentrations entering coastal water. Freshwater plumes from estuaries can extend hundreds of kilometres offshore (Figure 2.20) and the nutrients within them have a marked effect on patterns of primary productivity. Localized effects of eutrophication can be dramatic. For example, increased nitrogen supplies lead to the replacement of seagrass beds (e.g. *Zostera marina*) by free-floating rafts of ephemeral seaweeds such as *Ulva* and *Cladophera*, whose detritus may cover the bottom in a dense layer up to 50 cm thick.

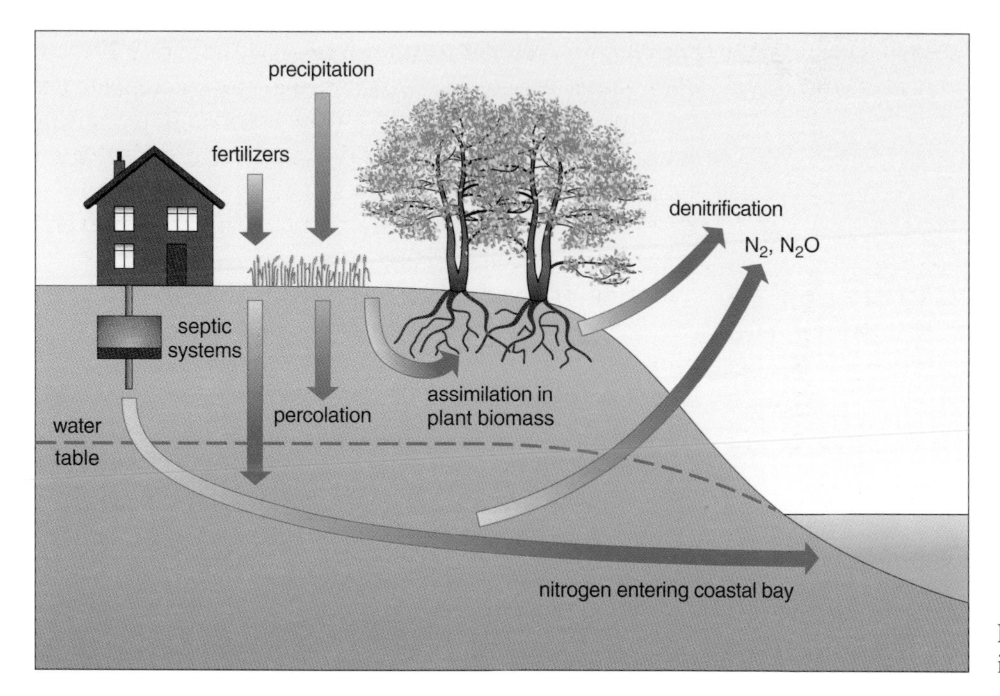

Figure 2.19 Nitrogen pathways in developed coastal estuaries.

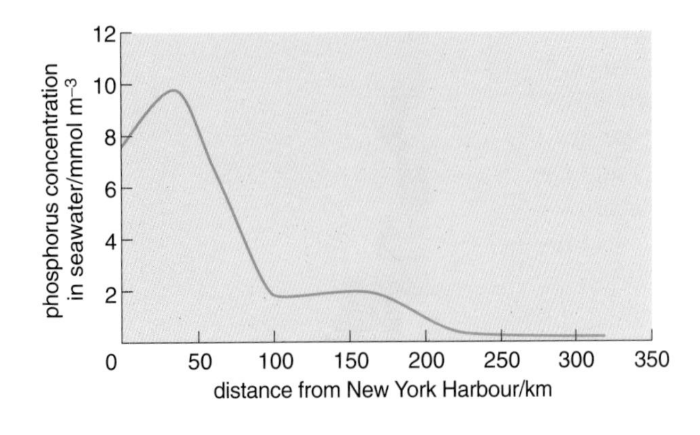

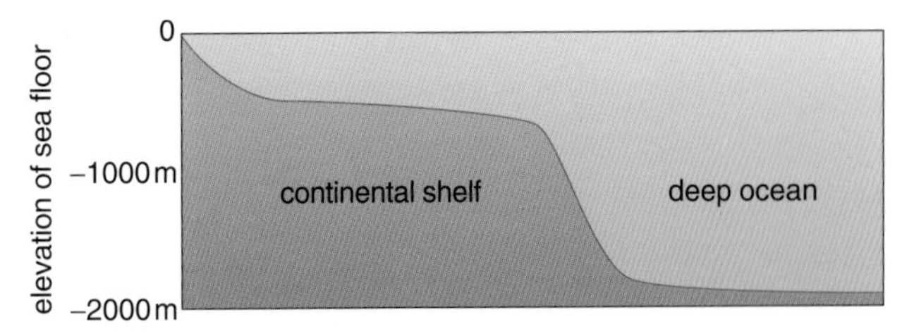

Figure 2.20 Phosphorus concentration with increasing distance from New York Harbour.

Estuarine waters enriched by nitrogen from fertilizers and sewage have been responsible for the decline of a number of estuarine invertebrate species, often by causing oxygen depletion of bottom water. Intertidal oyster beds have declined considerably as a result of both over-harvesting and reduced water quality. Harvesting tends to remove oysters selectively from shallow-water habitats, reducing the height of oyster beds and making remaining oysters more vulnerable to the damaging effects of eutrophication. In estuaries, elevated rates of microbial respiration deplete oxygen, and periods of anoxia occur more frequently, especially in summer when water temperatures are high and there is slow water circulation. Oysters in deeper water are more likely to be exposed to anoxic conditions, being further removed from atmospheric oxygen inputs, and to die as a result.

Seagrass distributions are very sensitive to variation in light. Seagrasses, like any other plant, cannot survive in the long term if their rate of photosynthesis is so limited by light that it cannot match their rate of respiration. Light transmission is a function of water column **turbidity** (cloudiness), which in turn is a combination of the abundance of planktonic organisms and the concentration of suspended sediment. Seagrass distributions are therefore strongly affected by eutrophication and effects on water clarity. In Chesapeake Bay, eastern USA, seagrass beds historically occurred at depths of over 10 m. Today they are restricted to a depth of less than 1 m. Runoff from organic fertilizers has increased plankton production in the water column, limiting light transmission. In addition, large areas of oyster bed have been lost (also partly as a result of eutrophication) with an associated reduction in natural filtration of bay water. Oysters, like many shellfish, clean the water while filtering out microbes, which they then consume. Seagrass beds now cover less than 10% of the area they covered a century ago (Figure 2.21). Similar patterns have occurred elsewhere.

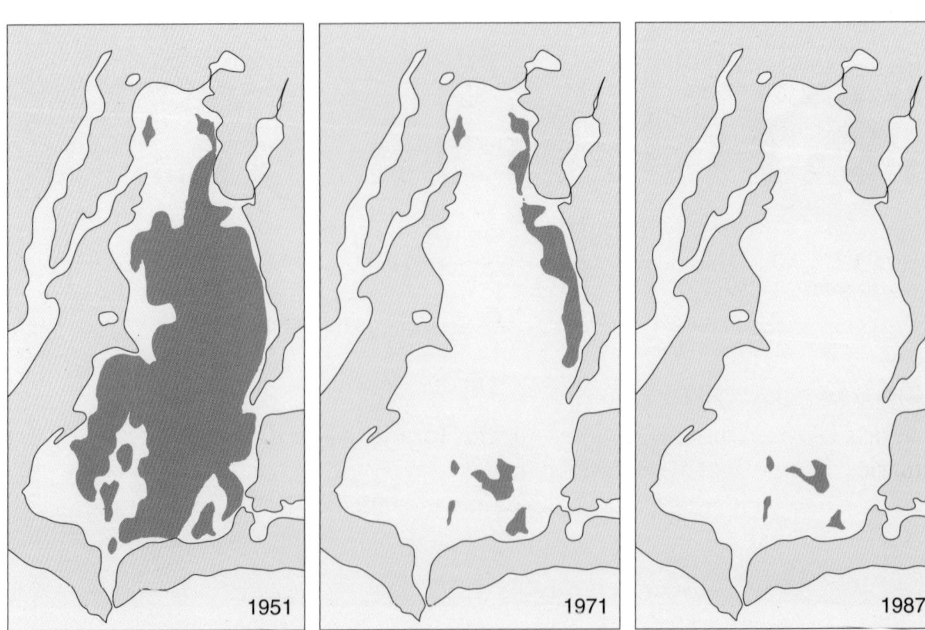

Figure 2.21 Decline of seagrass beds (green) in the estuary of Waquoit Bay, Cape Cod, over the past century.

Seagrass beds play an important role in reducing the turbidity of coastal waters by reducing the quantity of sediment suspended in the water. Seagrasses slow down currents near the bottom, which increases the deposition of small sediment particles and decreases their erosion and resuspension. Seagrass roots also play an important part in stabilizing sediments and limiting disturbance caused by burrowing deposit-feeders.

Many species depend on seagrass beds for food or nursery grounds. Seagrass increases the structural complexity of habitat near the sea floor, and provides a greater surface area for epiphytic organisms. Seagrass leaves support rich communities of organisms on their surface, including microalgae, stationary invertebrates (such as sponges and barnacles) and grazers (such as limpets and whelks). The plants also provide refuges from predatory fishes and crabs. Predation on seagrass-associated prey such as the grass shrimp is much higher outside seagrass beds than within them, where the shrimps can hide and predator foraging is inhibited by grass cover. Without seagrasses, soft-substrate communities on the sea bed are simpler, less heterogeneous and less diverse. By reducing the health of seagrasses, eutrophication contributes directly to biotic impoverishment (Figure 2.22).

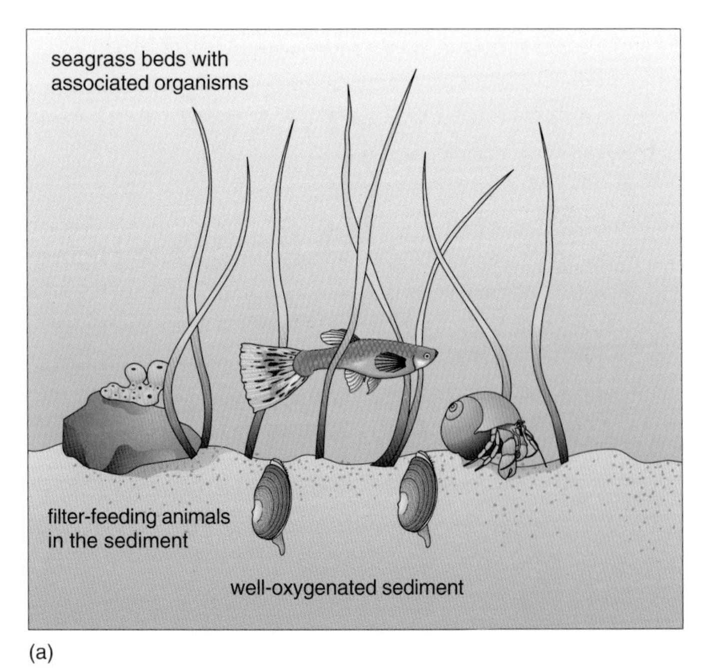

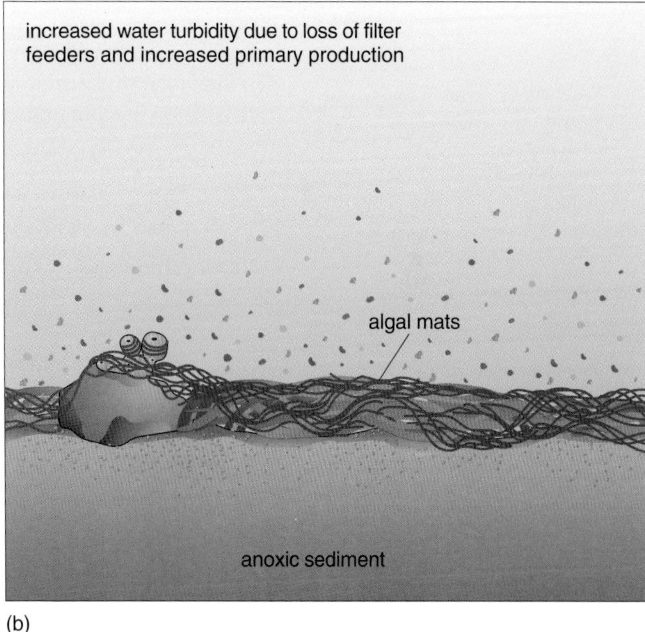

Figure 2.22 Changes in sea floor communities in shallow coastal waters following eutrophication. (a) The structural diversity afforded by the plants and the availability of oxygen in the sediment promote a diverse community of animals. (b) The loss of structural diversity and oxygen from the sea-bed causes the animal community to be replaced by one of bacterial decomposers.

Question 2.2

Seagrass (*Zostera marina*) is a key species for maintaining the biodiversity of estuaries. Why is its abundance reduced following eutrophication of estuarine waters? Can you identify a mechanism in which the decline of seagrass promotes its further decline?

2.4.2 Saltmarshes

Marsh plant primary production is generally nitrogen limited, so saltmarsh vegetation responds readily to the artificial eutrophication that is now so common in nearshore waters. Eutrophication causes marked changes in plant communities in saltmarshes, just as it does in freshwater aquatic and terrestrial systems. Biomass production increases markedly as levels of eutrophication increase. Increases in the nitrogen content of plants cause dramatic changes in populations of marsh plant consumers: insect herbivores tend to increase (Figure 2.23) and so do numbers of carnivorous insects. Thus, increasing the nitrogen supply to saltmarshes has a dramatic bottom-up effect on marsh food webs. Eutrophication can also alter the outcome of competition among marsh plants, by changing the factor limiting growth. At low levels of nitrogen, plants that exploit below-ground resources most effectively, such as the saltmarsh rush (*Juncus gerardii*) are competitively dominant, but at higher nutrient levels dominance switches to plants that are good above-ground competitors, such as the common cord grass (*Spartina anglica*, Figure 2.24). In other words, as nitrogen availability increases, competition for light becomes relatively more important.

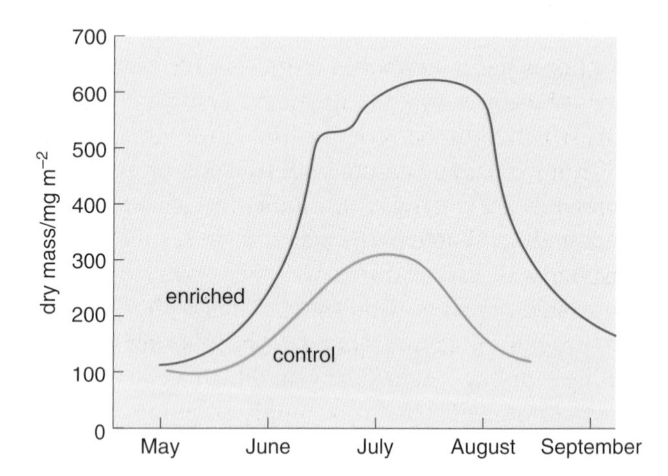

Figure 2.23 Effects of nutrient enrichment on herbivorous insect abundance (measured as dry mass) in saltmarshes.

Figure 2.24 The cord grass (*Spartina anglica*), which has spread rapidly around the coasts of Britain in the past 100 years, aided by the increased nutrient supply to saltmarshes in addition to being widely planted to help stabilize bare mud.

Causes and mechanisms of eutrophication

3

Light availability, water availability, temperature and the supply of plant nutrients are the four most important factors determining NPP. Altered availability of nutrients affects the rate of primary production in all ecosystems, which in turn changes the biomass and the species composition of communities.

3.1 Agents of eutrophication

- ❍ Which two elements most often limit NPP?
- ● Phosphorus and nitrogen are the main limiting nutrients.

Compounds containing these elements are therefore the causal agents of eutrophication in both aquatic and terrestrial systems. Let us consider them in turn.

3.1.1 Phosphorus

Phosphorus has a number of indispensable biochemical roles and is an essential element for growth in all organisms, being a component of nucleic acids such as DNA, which hold the code for life. However, phosphorus is a scarce element in the Earth's crust and natural mobilization of phosphorus from rocks is slow. Its compounds are relatively insoluble, there is no reservoir of gaseous phosphorus compounds available in the atmosphere (as there is for carbon and nitrogen), and phosphorus is also readily and rapidly transformed into insoluble forms that are unavailable to plants. This tends to make phosphorus generally unavailable for plant growth. In natural systems, phosphorus is more likely to be the growth-limiting nutrient than is nitrogen, which has a relatively rapid global cycle and whose compounds tend to be highly soluble.

Human activities, notably the mining of phosphate-rich rocks and their chemical transformation into fertilizer, have increased rates of mobilization of phosphorus enormously. A total of 12×10^{12} g P yr^{-1} are mined from rock deposits. This is six times the estimated rate at which phosphorus is locked up in the ocean sediments from which the rocks are formed. The global phosphorus cycle is therefore being unbalanced by human activities, with soils and water bodies becoming increasingly phosphorus-rich. Eutrophication produces changes in the concentrations of phosphorus in all compartments of the phosphorus cycle.

The mechanisms of eutrophication caused by phosphorus vary for terrestrial and aquatic systems. In soils, some phosphorus comes out of solution to form insoluble iron and aluminium compounds, which are then immobilized until the soil itself is moved by erosion. Eroded soil entering watercourses may release its phosphorus, especially under anoxic conditions.

- ❍ What changes occur to iron(III) compounds (Fe^{3+}) as a result of bacterial respiration in anoxic environments, and how is their solubility affected?
- ● Bacterial respiration can reduce Fe^{3+} to Fe^{2+}, increasing the solubility of iron salts, including phosphates of iron.

Once in rivers, retention times for phosphorus may be short, as it is carried downstream either in soluble form or as suspended sediment. Algal blooms are therefore less likely to occur in moving waters than in still systems. In the latter, there is more time for the phosphorus in enriched sediments to be released in an 'available' form, increasing the concentration of soluble reactive phosphorus (SRP), and thus affecting primary production.

Phosphorus is generally acknowledged to be the nutrient most likely to limit phytoplankton biomass, and therefore also the one most likely to cause phytoplankton blooms if levels increase. However, there do appear to be some systems that are 'naturally eutrophic', with high phosphorus loadings. In these systems, nitrogen concentrations may then become limiting and play a dominant role in determining phytoplankton biomass.

3.1.2 Nitrogen

Nearly 80% of the atmosphere is nitrogen. Despite the huge supply potentially available, nitrogen gas is directly available as a nutrient to only a few organisms.

- ❍ Why cannot the majority of organisms utilize gaseous nitrogen?
- ● Nitrogen gas is very unreactive and only a limited number of bacterial species have evolved an enzyme capable of cleaving the molecule.

Once 'fixed' by these bacteria into an organic form, the nitrogen enters the active part of the nitrogen cycle. As the bacteria or the tissues of their mutualistic hosts die, the nitrogen is released in an available form such as nitrate or ammonium ions – a result of the decay process. Alternatively, the high temperatures generated during electrical storms can 'fix' atmospheric nitrogen as nitric oxide (NO). Further oxidation to nitric acid within the atmosphere, and scavenging by rainfall, provides an additional natural source of nitrate to terrestrial ecosystems. Nitrates and ammonium compounds are very soluble and are hence readily transported into waterways.

Nitrogen is only likely to become the main growth-limiting nutrient in aquatic systems where rocks are particularly phosphate-rich or where artificial phosphate enrichment has occurred. However, nitrogen is more likely to be the limiting nutrient in terrestrial ecosystems, where soils can typically retain phosphorus while nitrogen is leached away.

3.2 Anthropogenic sources of nutrients

In addition to the natural sources of nutrients referred to above, nitrogen and phosphorus enter the environment from a number of anthropogenic sources. These are considered below.

3.2.1 The atmosphere

Pollution of the atmosphere has increased rates of nitrogen deposition considerably. Nitrogen has long been recognized as the most commonly limiting nutrient for terrestrial plant production throughout the world, but air pollution has now created a modern, chemical, climate that often results in excess supplies of nitrogen due to atmospheric deposition.

The main anthropogenic source of this enhanced nitrogen deposition is the NO_x (mainly as NO) released during the combustion of fossil fuels — principally in vehicles and power plants. Like that generated within the atmosphere, this fixed nitrogen returns to the ground as nitrate dissolved in rainwater.

Figure 3.1 Industrial emissions. These introduce nitrogen compounds into the atmosphere.

Patterns and rates of deposition vary regionally, and between urban and rural areas. Concentrations and fluxes of nitrogen oxides tend to decline with distance from cities: deposition of inorganic nitrogen has been found to be twice as high in urban recording sites in New York City than in suburban or rural sites. Some natural ecosystems, particularly those near industrialized areas, now receive atmospheric nitrogen inputs that are an order of magnitude greater than those for pre-industrial times. Figure 3.2 shows intensive industrial land use adjacent to the River Tees and its estuary in Teesside, UK. The estuary is still important for wildlife, including seals and a variety of birds, but its quality has declined markedly due to atmospheric and water pollution. In the UK, atmospheric deposition can add up to 150 kg N ha^{-1} yr^{-1}. For comparison, the amount thought to trigger changes in the composition of species-rich grassland is 20–30 kg N ha^{-1} yr^{-1}, and a typical dose farmers apply as inorganic fertiliser to an intensively managed grassland is 100 kg N ha^{-1} yr^{-1}.

Figure 3.2 Large-scale industrial development adjacent to the River Tees.

3.2.2 Domestic detergents

Domestic detergents are a major source of phosphorus in sewage effluents. Phosphates are used as a 'builder' in washing powders to enhance the efficiency of surfactants by removing calcium and magnesium to make the water 'softer' (see Topic 6 *Water Quality*). In 1992, the UK used 845 600 tonnes of detergent of various types, all of which have different effects on the environment. Estimates of the relative contribution of domestic detergents to phosphorus build-up in Britain's watercourses vary from 20–60%. The UK's Royal Commission on Environmental Pollution (RCEP) reviewed the impacts of phosphate-based detergents on water quality in 1992, focusing on the effects on freshwater. The RCEP concluded that eutrophication was widespread over large parts of the country, and recommended a considerable investment in stripping phosphates from sewage as well as efforts to reduce phosphate use in soft-water areas. The main problem is that many of the ingredients of detergents are not removed by conventional sewage treatment and degrade only slowly.

❍ Why did the RCEP recommend that phosphate use be reduced particularly in 'soft' water areas?

● The reason for including phosphates in detergents is to soften the water, so in areas with naturally soft water they provide no benefit yet still cause pollution.

Other compounds added to detergents may also contribute to eutrophication. Silicates, for example, particularly if used as a partial replacement for phosphates in detergents, can lead to increased growth of diatoms. These algae require silicates to build their 'skeleton' and their growth can be limited by silicate availability. When silicates are readily available, diatoms characteristically have 'spring blooms' of rapid growth, and can smother the surfaces of submerged macrophytes, depriving them of light. A loss of submerged macrophytes is a problem because it results in the loss of habitat for organisms feeding on phytoplankton, and therefore the risk of blooms by *other* species is enhanced.

3.2.3 Agricultural fertilizers

Runoff from intensively farmed land often contains high concentrations of inorganic fertilizer. Nutrients applied to farmland may spread to the wider environment by:

- drainage water percolating through the soil, leaching soluble plant nutrients;
- washing of excreta, applied to the land as fertilizer, into watercourses; and
- the erosion of surface soils or the movement of fine soil particles into subsoil drainage systems.

Some water bodies have been monitored for long periods, and the impact of agricultural runoff can be demonstrated clearly. In the 50 years between 1904 and 1954, for example, in Loch Leven, Scotland, there were major changes in the species composition of the community of photosynthetic organisms. The species composition of the green alga community changed and the numbers of cyanobacteria rose considerably. Increasingly since then, large blooms of filamentous cyanobacteria have been produced in the loch. These changes have been linked with trends in the use of agricultural fertilizers and other agrochemicals.

In Europe, large quantities of slurry from intensively reared and housed livestock are spread on the fields (Figure 3.3). Animal excreta are very rich in both nitrogen and phosphorus and therefore their application to land can contribute to problems from polluted runoff. Land use policies have concentrated livestock production into purpose-built units, increasing the pollution risks associated with handling the resultant slurry or manures.

European agricultural policies that subsidize agriculture on the basis of productivity have also encouraged the use of fertilizers. Use of fertilizers has undergone a massive increase since 1950. In the USA, by 1975, total use of inorganic fertilizer had reached a level equivalent to about 40 kg per person per year. A recent European Environment Agency report estimated that the groundwater beneath more than 85% of Europe's farmland exceeds guideline levels for nitrogen concentration (25 mg l^{-1}), with agricultural fertilizers being the main source of the problem. Pollution of surface waters also occurs on a large scale. A survey by the UK's Environment Agency in 1994 found that over 50% of the 314 water bodies surveyed in England and Wales had algal blooms caused by fertilizer runoff (Figure 3.4).

Figure 3.3 Muck spreading.

Patterns of fertilizer use do differ considerably between countries. In those with poorly developed economies, the costs of artificial fertilizers may be prohibitive. In hotter climates, irrigation may be used, resulting in higher nutrient runoff than for equivalent crops that are not irrigated. The high solubility of nitrate means that agriculture is a major contributor to nitrogen loadings in freshwater. Agriculture accounts for 71% of the mass flow of nitrogen in the River Great Ouse in the Midlands, UK, compared with only 6% for phosphorus.

Figure 3.4 Algal blooms caused by fertilizer runoff.

- ❍ What are the main sources of phosphorus and nitrogen that enrich rivers in a developed country such as the UK?
- ● Phosphorus comes primarily from domestic wastewater, whilst nitrogen comes primarily from intensive agriculture.

3.2.4 Land use

Studies evaluating the effects of nutrient loading on receiving water bodies must take account of the range of land uses found within a catchment.

As shown in Table 3.1, phosphate exports increase considerably as forests are converted to agricultural land and as agricultural land is urbanized. Agricultural runoff is known to be a potential source of nutrients for eutrophication, but the degree of mechanization may also be important. In catchments where agriculture is heavily mechanized, higher levels of sedimentation are likely. Most sediments arise as a result of soil erosion, which is promoted by tilling the land intensively. This destroys the soil's natural structure as well as removing vegetation which helps to stabilize soil.

To cite just one example, high sediment input in the latter half of the 20th century has caused shrinkage of the area of open water in the Mogan Lake system near Ankara, Turkey. Undoubtedly, mechanization and intensification of agriculture have played their part, but so too has the drainage of adjacent wetlands. The drained wetlands no longer trapped sediments, and themselves became vulnerable to erosion. This further increased sediment loadings in the lake. Levels of phosphorus have also risen. Draining the wetlands exposed the organic matter in their soils to oxidation, 'mobilizing' the phosphorus that had accumulated there over many years. This was then carried into the lake in drainage water.

Table 3.1 Quantities of nitrogen and phosphorus (g m^{-2} yr^{-1}) derived from various types of land use in the USA and from the atmosphere.

	Land use	Total phosphorus	Total nitrogen
Losses from land to water courses	urban	0.1	0.5
	rural/agriculture	0.05	0.5
	forest	0.01	0.3
Additions to land	atmospheric sources:		
	rainfall	0.02	0.8
	dry deposition	0.08	1.6

3.2.5 Sediments

Sediments have a variable but complex role in nutrient cycling in most aquatic systems, and are a potential 'internal' source of pollutants. Release of phosphorus from lake sediment is a complex function of physical, biological and chemical processes and is not easy to predict for different systems. Nitrogen is not stored and released from sediments in the same way, so its turnover time within aquatic systems is quite rapid. Nitrogen concentrations tend to fall off relatively quickly following a reduction in external nitrogen loading, whilst this is not true for phosphorus because the sediments can hold such a large reservoir of this nutrient that input and output rates may become decoupled.

In some shallow coastal areas, tidal mixing is the dominant nutrient regeneration process, as the sediments are regularly disturbed and redistributed by changing water currents, making nutrient exchange with the water much more rapid.

- ❍ Why is a lake in a catchment dominated by arable agriculture much more prone to eutrophication than one in a forested catchment?
- ● First the arable catchment is likely to be receiving much more nutrient input in the form of fertilizers. Secondly and equally importantly, the soil structure is much less stable under arable systems and therefore more likely to erode and carry nutrients to the lake as suspended sediment.

3.3 Mechanisms of eutrophication

Direct effects of eutrophication occur when growth of organisms (usually the primary producers) is released from nutrient limitation. The resulting increased NPP becomes available for consumers, either as living biomass for herbivores or as detritus for detritivores. Associated indirect effects occur as eutrophication alters the food supply for other consumers. Changes in the amount, relative abundance, size or nutritional content of the food supply influence competitive relationships between consumers, and hence the relative success and survival of different species. Nutrient-induced changes in plant community composition and productivity can therefore result in associated changes in the competitive balance between herbivores, detritivores and predators. Consumers may also be affected by changes in environmental conditions caused indirectly by eutrophication, for example reduced oxygen concentrations caused by bacterial decay of biomass.

In freshwater aquatic systems, a major effect of eutrophication is the loss of the submerged macrophyte community (Section 2.1.1). Macrophytes are thought to disappear because they lose their energy supply in the form of sunlight penetrating the water. Following eutrophication, the sunlight is intercepted by the increased biomass of phytoplankton exploiting the high availability of nutrients. In principle, the submerged macrophytes could also benefit from increased nutrient availability, but they have no opportunity to do so because they are shaded by the free-floating microscopic organisms. Research in the Norfolk Broads has supported the view that the rapid replacement of diverse macrophyte communities by algal communities is attributable to light attenuation, caused by raised turbidity, but has also suggested that there may be more complex mechanisms operating, which must be understood if practical measures are to be undertaken to tackle eutrophication problems. There is evidence to suggest that *either* a plant-dominated state *or* an algal-dominated state can exist under high-nutrient conditions (Figure 3.5). Once either state becomes established, a number of mechanisms come into play which buffer the ecosystem against externally applied change. For example, a well-established submerged plant community may secrete substances that inhibit algal growth, and may provide refuges for animals that graze large quantities of algae. On the other hand, once an algal community becomes well established, especially early in the year, it can shade out the new growth of any aquatic plants on the bottom and compete with them for carbon dioxide in the water.

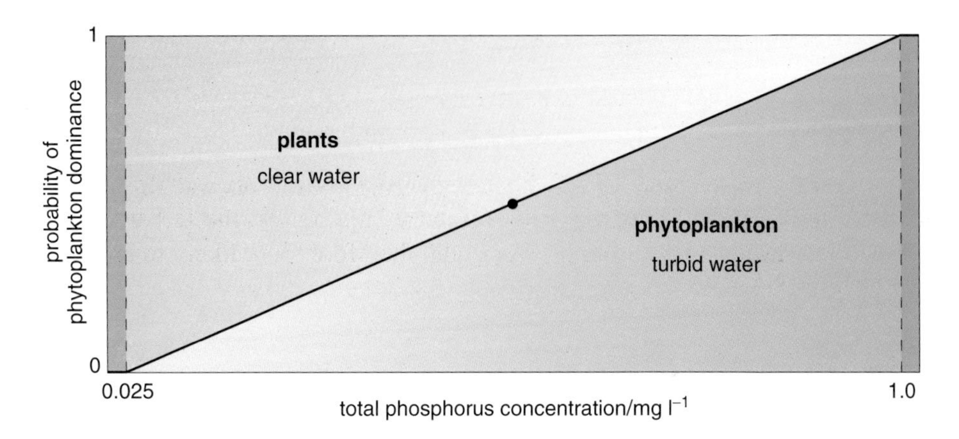

Figure 3.5 Probability plot of two stable states in shallow freshwater ecosystems. Over a broad range of phosphorus concentrations in the eutrophic–hypertrophic range, either state may potentially occur. However, once established, that state promotes processes that result in it becoming stabilized, and switches between the two states are only rarely observed.

Research in the Norfolk Broads into possible trigger factors for switches from communities dominated by submerged macrophytes to those dominated by algae suggests that pesticides could play a role. Some herbivores are thought to be susceptible to pesticide leaching from surrounding arable land. Pesticide residues in sediments were found at concentrations high enough to cause at least sub-lethal effects, which could reduce the herbivore population for long enough to reduce algal consumption. This could help to explain the observation that most of the Norfolk Broads that have lost their plants are directly connected with main rivers draining intensive arable catchments, whereas those that have retained plant dominance are in catchments where livestock grazing predominates.

Clear relationships can be seen between human population density and total phosphorus and nitrate concentrations in watercourses (Figure 3.6). In 1968 the anthropogenic contribution amounted to some 10.8 g N per capita per day and 2.18 g P per capita per day. Outputs have continued to rise since then. Worldwide, human activities have intensified releases of phosphorus considerably. Increased soil erosion, agricultural runoff, recycling of crop residues and manures, discharges of domestic and industrial wastes and, above all, applications of inorganic fertilizers, are the major causes of this increase. Global food production is now highly dependent on the continuing use of supplementary phosphates, which account for 50–60% of total phosphorus supply.

Studies of nutrient runoff have shown a mixture of inputs into most river and lake catchments: both **point source** (such as sewage treatment works) and **diffuse source** (such as agriculture). Point sources are usually most important in the supply of phosphorus, whereas nitrogen is more likely to be derived from diffuse sources.

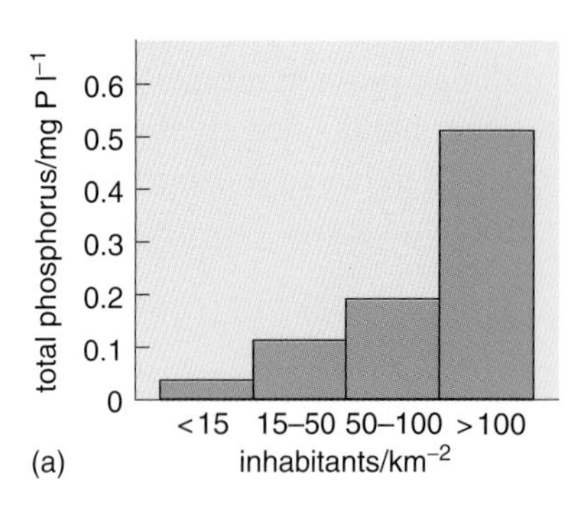

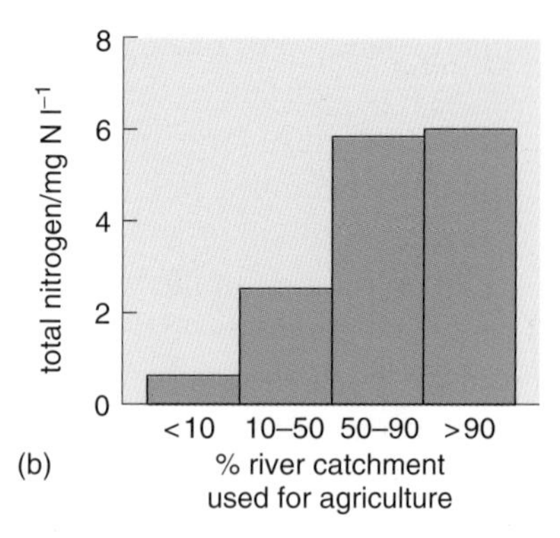

Figure 3.6 (a) The relationship between human population density and the concentration of phosphorus (P) in rivers across Europe. (b) The relationship between the percentage of catchment in agriculture and the concentration of nitrogen (N) in the same European rivers.

Question 3.1

Using the data presented in Figure 1.13 and Table 2.3, comment on whether the remediation activities on the broads neighbouring the River Ant were likely to have resulted in a recovery of plant species diversity by 2000. Assume that 80% of the total phosphorus in the water is in the form of SRP.

Managing eutrophication

4

The degree to which eutrophication is considered a problem depends on the place and people concerned. A small lake in South-East Asia, heavily fertilized by village sewage, can provide valuable protein from fish. In other parts of the world, a similar level of nutrients would be regarded as damaging, making water undrinkable and unable to support characteristic wildlife. In Europe, nitrates in drinking water are regarded as a potentially serious threat to health. Eutrophication has also damaged important fisheries and caused significant loss of biodiversity. Worldwide, efforts to reduce the causes and symptoms of eutrophication cost huge sums of money.

There is no single piece of existing legislation dealing comprehensively with the problem of eutrophication in the UK. However, one aim of the European Community's Urban Wastewater Treatment Directive (EC UWWTD) is to protect the environment from the adverse effects of sewage. This should help to reduce the problem of eutrophication in coastal waters where large discharges

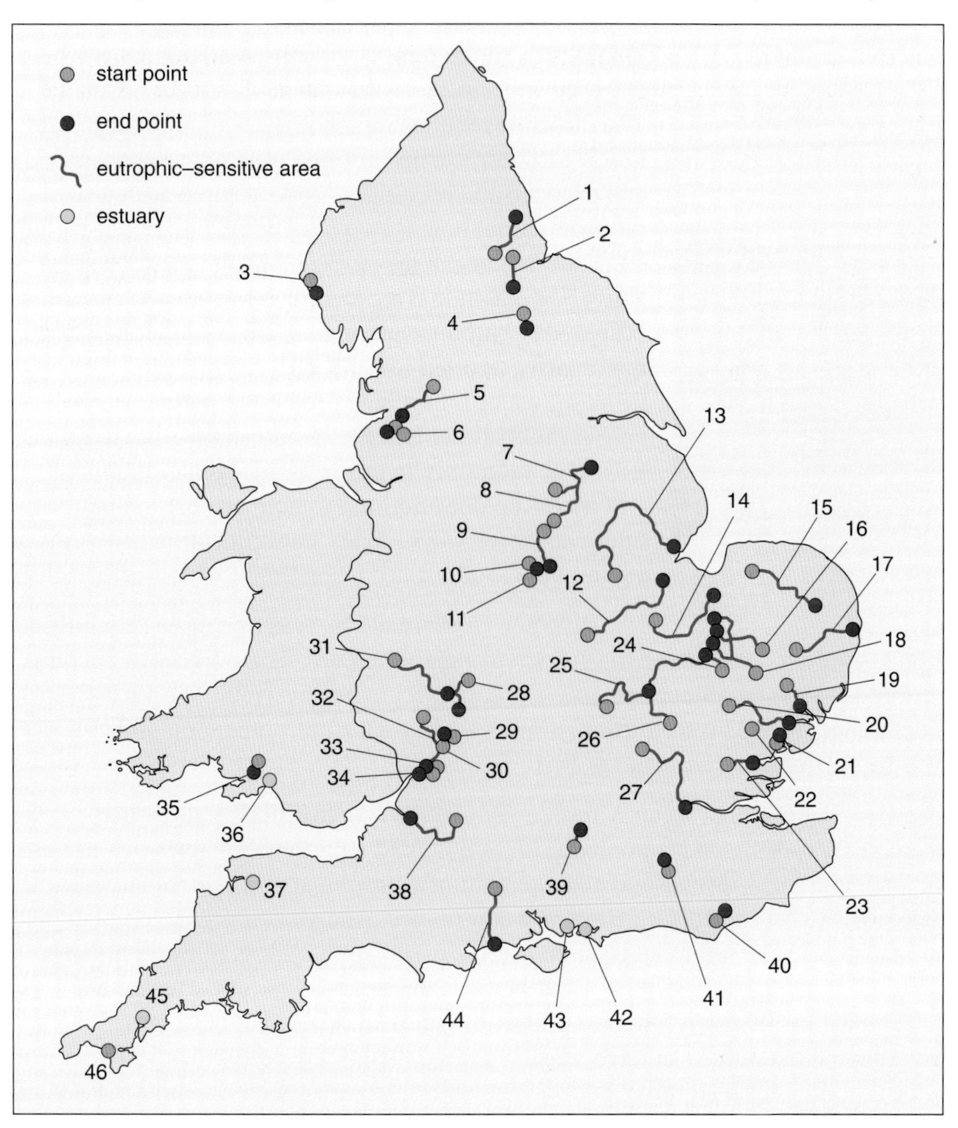

1 River Wear
2 River Skerne
3 River Ehen
4 River Wiske
5 Lower River Ribble
6 River Yarrow
7 Chesterfield Canal
8 River Idle catchment
9 River Erewash
10 River Derwent
11 Staunton Harold reservoirs
12 River Welland
13 River Witham
14 Middle Level
15 River Wensum
16 Little Ouse
17 River Waveney
18 River Lark
19 River Gipping
20 River Stour
21 Abbertson Reservoir
22 River Colne
23 Rivers Chelmer Wid and Can
24 Soham Lode
25 River Great Ouse
26 River Ivel
27 River Lee & Lee Navigation
28 Bow Brook
29 River Chelt
30 Gloucester Sharpness Canal
31 River Teme
32 River Leadon
33 River Frome
34 River Cam
35 River Lougher
36 Tawe
37 Taw Estuary
38 Bristol Avon
39 River Loddon
40 Pevensey Levels
41 River Mole and Gatwick Stream
42 Chichester Harbour
43 Langstone Harbour
44 River Avon (Hampshire)
45 Fal
46 Loe Pool

Figure 4.1 Map of eutrophic-sensitive waters in the UK, designated by European Directive.

contribute significant nutrient loads. In the UK, 62 rivers and canals (totalling 2500 km), 13 lakes and reservoirs and five estuaries have been designated as sensitive areas (eutrophic) under this directive, and there are requirements for reducing nutrient loads from sewage treatment works in these areas (Figure 4.1).

Figure 4.2 Sewage treatment works.

Under the EC UWWTD, areas designated as 'eutrophic sensitive' must have phosphorus-stripping equipment installed at sewage treatment works (STWs, Figure 4.2) that serve populations of 10 000 or more. However, the majority of nature conservation sites classified as sensitive are affected by smaller, rural STWs for which such equipment is not yet required. Phosphorus stripping involves the use of chemicals such as aluminium sulfate, which react with dissolved phosphates, causing them to precipitate out of solution.

Figure 4.3 Slurry handling.

Another piece of European legislation that has some bearing on problems of nutrient enrichment is the EC Nitrates Directive. This is intended to reduce nitrate loadings to agricultural land, particularly in areas where drinking water supplies have high dissolved nitrate levels. The directive requires member states to monitor nitrate levels in water, set up 'nitrate vulnerable zones' (NVZ), and produce and promote a 'code of good agricultural practice' throughout the countryside. This should include measures to control the storage, handling and disposal of slurry, for example (Figure 4.3). However, the legislation designed to curb nutrient inputs from agricultural sources is primarily directed towards reducing nitrate levels in drinking water rather than protecting nature conservation sites. The EC Nitrates Directive defines eutrophication only in terms of nitrogen compounds, and therefore does nothing to help protect the majority of aquatic sites where many eutrophication problems are attributable to phosphorus loading.

The Declaration of the Third North Sea Conference in 1990 specified that nutrient inputs entering areas of the marine environment that are, or are likely to become, eutrophic, must be reduced to 50% of their 1985 levels by 1995. The Fifth Conference in 2002 went further, aiming to eliminate eutrophication and create a healthy marine environment by 2010. Fine words.

The UK Environment Agency has developed a eutrophication strategy that promotes a coordinated framework for action, and a partnership approach at both national and local levels. The management of eutrophication requires targets and objectives to be agreed for different water bodies. Analysis of preserved plant and animal remains in sediments can be used to estimate the levels of nutrients that occurred in the past, when the water bodies concerned were less affected by eutrophication. These reference conditions can then be used to determine which waters are most at risk, or have already been damaged by eutrophication, and to prioritize sites for restorative action. The ability to measure and monitor levels of eutrophication has therefore become increasingly important.

4.1 Measuring and monitoring eutrophication

During the 1990s there was increased demand in the UK for effective methods of monitoring eutrophication. There was also considerable interest in the development of monitoring systems based on biotic indices. Several 'quality indices' based on a variety of organisms were explored. For monitoring tools to have practical application, they must satisfy certain requirements:

- sampling must be quick and easy;
- monitoring must be based on a finite number of easily identified groups; and
- indices for evaluation must be straightforward to calculate.

❍ In your tour of the Teign Catchment (Block 1), what system of monitoring using a biotic index did you encounter?

● The BMWP system for evaluating water quality based on the presence or absence of invertebrate animal species.

Within-year variability in nutrient concentrations can be high, particularly for enriched waters. A high sampling frequency may therefore be required to provide representative annual mean data. In nutrient-enriched lakes, annual means are more likely to provide appropriate estimates of phosphorus than winter–spring means, due to the importance of internal cycling of nutrients in summer. This is an important consideration when designing sampling strategies for use in predictive models of trophic status.

The large group of algal species collectively known as diatoms has been used as indicators of eutrophication in European rivers. Individual species of diatom vary in their tolerance of nutrient enrichment, some species being able to increase their growth rates as nutrients become more available, whilst others are outcompeted and disappear. As diatoms derive their nutrients directly from the water column, and have generation times measured in days rather than months or years, the species composition of the diatom community should be a good indicator for assessing eutrophication. Convincing correlations have been demonstrated between aqueous nutrient concentrations and diatom community composition, but there are a number of other physical and chemical factors that also affect diatom distribution, such as water pH, salinity and temperature, which also need to be taken into account.

The UK Environment Agency has assessed the extent of eutrophication on the basis of concentrations of key nutrients (primarily nitrogen and phosphorus) in water, and the occurrence of obvious biological responses, such as algal blooms. There is an intention to rely more heavily in future on biological assessment schemes. One such system is based on surveys of the aquatic plant populations in rivers. Known as the **mean trophic rank (MTR)** approach, this uses a scoring system based on species and their recorded abundances at river sites. Each species is allocated a score (its **species trophic rank, STR**) dependent on its tolerance to eutrophication (Table 4.1); then, for a given site, the mean score for all species present is calculated. Tolerant species have a low score, so a low MTR tends to indicate a nutrient-rich river. In Britain, rivers in the north and west tend to have the highest MTR scores, whereas rivers in the south and east of England have the lowest. These scores reflect the influence of numerous factors, such as differences in river flow, patterns of agricultural intensification and variations in population density.

Table 4.1 Sensitivity of aquatic plants to nutrient enrichment, as indicated by species trophic rank (STR).

Species	STR	Species	STR	Species	STR
Algae		**Angiosperms**		**Angiosperms**	
Batrachospermum spp.	6	**(a) Broadleaved species**		**(b) Grassleaved species**	
Hildenbrandia rivularis	6	*Apium inundatum*	9	*Acorus calamus*	2
Lemanea fluviatilis	7	*A. nodiflorum*	4	*Alisma plantago-aquatica*	3
Vaucheria spp.	1	*Berula erecta*	5	*A. lanceolatum*	3
Cladophora spp.	1	*Callitriche hamulata*	9	*Butomus umbellatus*	5
Enteromorpha spp.	1	*C. obtusangula*	5	*Carex acuta*	5
Hydrodictyum reticulatum	3	*Ceratophyllum demersum*	2	*C. acutiformis*	3
Stigeoclonium tenue	1	*Hippurus vulgaris*	4	*C. riparia*	4
		Littorella uniflora	8	*C. rostrata*	7
Liverworts		*Lotus pedunculatus*	8	*C. vesicaria*	6
Chiloscyphus polyanthos	8	*Menyanthes trifoliata*	9	*Catabrosa aquatica*	5
Jungermannia atrovirens	8	*Montia fontana*	8	*Eleocharis palustris*	6
Marsupella emarginata	10	*Myriophyllum alterniflorum*	8	*Eleogiton fluitans*	10
Nardia compressa	10	*M. spicatum*	3	*Elodea canadensis*	5
Pellia endiviifolia	6	*Myriophyllum* spp.*	6	*E. nuttallii*	3
P. epiphylla	7	*Nuphar lutea*	3	*Glyceria maxima*	3
Scapania undulata	9	*Nymphaea alba*	6	*Groenlandia densa*	3
		Nymphoides peltata	2	*Hydrocharis morsus-ranae*	6
Mosses		*Oenanthe crocata*	7	*Iris pseudacorus*	5
Amblystegium fluviatilis	5	*O. fluviatilis*	5	*Juncus bulbosus*	10
A. riparium	1	*Polygonum amphibium*	4	*Lemna gibba*	2
Blindia acuta	10	*Potentilla erecta*	9	*L. minor*	4
Brachythecium plumosum	9	*Ranunculus aquatilis*	5	*L. minuta/miniscula*	3
B. rivulare	8	*R. circinatus*	4	*L. trisulca*	4
B. rutabulum	3	*R. flammula*	7	*Phragmites australis*	4
Bryum pseudotriquetrum	9	*R. fluitans*	7	*Potamogeton alpinus*	7
Calliergon cuspidatum	8	*R. omiophyllus*	8	*P. berchtoldii*	4
Cinclidotus fontinaloides	5	*R. peltatus*	4	*P. crispus*	3
Dichodontium flavescens	9	*R. penicillatus pseudofluitans*	5	*P. friesii*	3
D. pellucidum	9	*R. penicillatus penicillatus*	6	*P. gramineus*	7
Dicranella palustris	10	*R. penicillatus vertumnus*	5	*P. lucens*	3
Fontinalis antipyretica	5	*R. trichophyllus*	6	*P natans*	5
F. squamosa	8	*R. hederaceus*	6	*P. obtusifolia*	5
Hygrohypnum luridum	9	*R. sceleratus*	2	*P. pectinatus*	1
H. ochraceum	9	*Ranunculus* spp.*	6	*P. perfoliatus*	4
Hyocomium armoricum	10	*Rorippa amphibia*	3	*P. polygonifolius*	10
Philonotis fontana	9	*R. nasturtium-aquaticum*	5	*P. praelongus*	6
Polytrichum commune	10	*Rumex hydrolapathum*	3	*P. pusillus*	4
Racomitrium aciculare	10	*Veronica anagallis-aquatica*	4	*P. trichoides*	2
Rhynchostegium riparioides	5	*V. catenata*	5	*Sagittaria sagittifolia*	3
Sphagnum spp.	10	*V. scutellata*	7	*Schoenoplectus lacustris*	3
Thamnobryum alopecurum	7	*Viola palustris*	9	*Scirpus maritimus*	3
				Sparganium emersum	3
Fern-allies				*S. erectum*	3
Azolla filiculoides	3			*Spirodela polyrhiza*	2
Equisetum fluviatile	5			*Typha latifolia*	2
E. palustre	5			*T. angustifolia*	2
				Zannichellia palustris	2

Response to eutrophication:

STR 1–3 most tolerant   STR 4–5 moderately tolerant   STR 6–7 moderately sensitive   STR 8–10 most sensitive

* Average values for the genus are used when individual species cannot be identified.

4.2 Reducing eutrophication

In Britain, water supply companies have tended to regard eutrophication as a serious problem only when it becomes impossible to treat drinking water supplies in an economic way. Threshold concentrations at which action is taken to reduce nutrient loadings thus depend on economic factors, as well as wildlife conservation objectives.

There are two possible approaches to reducing eutrophication:

1 Reduce the source of nutrients (e.g. by phosphate stripping at sewage treatment works, reducing fertilizer inputs, introducing buffer strips of vegetation adjacent to water bodies to trap eroding soil particles).
2 Reduce the availability of nutrients currently in the system (e.g. by removing plant material, removing enriched sediments, chemical treatment of water).

4.3 Reducing the nutrient source

Europe is the continent that has suffered most from eutrophication, and increasing efforts are being made to restore European water bodies damaged by nutrient enrichment. If the ultimate goal is to restore sites where nature conservation interest has been damaged by eutrophication, techniques are required for reducing external loadings of nutrients into ecosystems.

Although algal production requires both nitrogen and phosphorus supplies, it is usually sufficient to reduce only one major nutrient. An analogy can be drawn with motor cars, which require lubricating oil, fuel and coolant to keep them moving and are likely to stop if they run short of any one of these, even if the other two are in plentiful supply. As phosphorus is the limiting nutrient in most freshwater systems, phosphorus has been the focus of particular attention in attempts to reduce inputs. In addition, nitrogen is less easily controlled: its compounds are highly soluble and can enter waterways from many diffuse sources. It can also be 'fixed' directly from the atmosphere. Phosphorus, on the other hand, is readily precipitated, usually enters water bodies from relatively few point sources (e.g. large livestock units or waste-water treatment works) and has no atmospheric reserve. However, efforts to reduce phosphorus loadings in some lakes have failed due to ongoing release of phosphorus from sediments. In situations where phosphorus has accumulated naturally (e.g. in areas with phosphate-rich rocks) and nitrogen increases have driven eutrophication, it may be necessary to control nitrogen instead.

4.3.1 Diversion of effluent

In some circumstances it may be possible to divert sewage effluent away from a water body in order to reduce nutrient loads. This was achieved at Lake Washington, near Seattle, USA, which is close to the sea. Lake Washington is surrounded by Seattle and its suburbs, and in 1955 a cyanobacterium, *Oscitilloria rubescens*, became dominant in the lake. The lake was receiving sewage effluent from about 70 000 people; this input represented about 56% of the total phosphorus load to the lake. The sewerage system was redesigned to divert effluent away from the lake, for discharge instead into the nearby sea inlet of Puget Sound. The transparency of the water in the lake, as measured by the depth at which a white disc could be seen, quickly increased from about 1 to 3 m, and chlorophyll concentrations decreased markedly as a result of reduced bacterial populations.

Diversion of effluent should be considered only if the effluent to be diverted does not constitute a major part of the water supply for the water body. Otherwise, residence times of water and nutrients will be increased and the benefits of diversion may be counteracted.

4.3.2 Phosphate stripping

It has been estimated that up to 45% of total phosphorus loadings to freshwater in the UK comes from sewage treatment works. This input can be reduced significantly (by 90% or more) by carrying out phosphate stripping. The effluent is run into a tank and dosed with a product known as a precipitant, which combines with phosphate in solution to create a solid, which then settles out and can be removed. It is possible to use aluminium salts as a precipitant, but the resulting sludge contains toxic aluminium compounds that preclude its secondary use as an agricultural fertilizer. There are no such problems with iron salts, so Fe(II) ammonium sulfate is frequently chosen as a precipitant. The chemicals required as precipitants constitute the major cost, rather than installations or infrastructure, and the process is very effective: up to 95% of the phosphate can be removed easily, and it is possible to remove more. Despite its effectiveness, however, phosphate stripping is not yet used universally in sewage treatment.

4.3.3 Buffer strips

The interface between aquatic ecosystems and the land is an ecotone that has a profound influence on the movement of water and water-borne contaminants. Vegetation adjacent to streams and water bodies can help to safeguard water quality, particularly in agricultural landscapes. **Buffer strips** are used to reduce the amounts of nutrients reaching water bodies from runoff or leaching. They usually take the form of vegetated strips of land alongside water bodies: grassland, woodland and wetlands have been shown to be effective in different situations. The vegetation often performs a dual role, by reducing nutrient inputs to aquatic habitat and also providing wildlife habitat. A riparian buffer zone of between 20 and 30 m width can remove up to 100% of incoming nitrate. The plants take up nitrogen directly, provide a source of carbon for denitrifying bacteria and also create oxidized **rhizospheres** where denitrification can occur. Uptake of nitrogen by vegetation is often seasonal and is usually greater in forested areas with sub-surface water flow

(a)

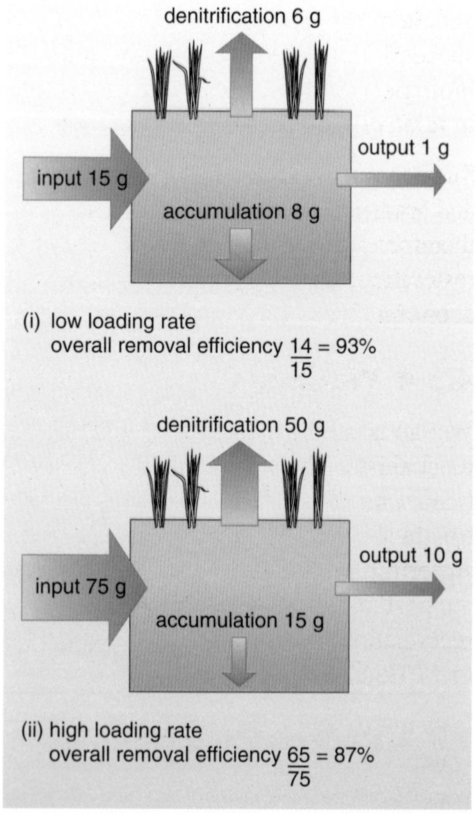

(b)

Figure 4.4 (a) A field experiment investigating the effectiveness with which a grass buffer strip prevents nutrients applied to the arable field beyond from reaching the stream. In the foreground the V-notch weir is being used to monitor flow, and the small shed houses automatic equipment to sample and assess water quality at regular intervals. (b) Results from such experiments indicate that the majority of the dissolved nitrogen entering the buffer strip is either retained or volatilized, with little reaching the stream even at relatively high loading rates.

than in grassland with predominantly surface flow. The balance between surface flow and sub-surface flow, and the redox conditions that result, are critical in determining rates of nitrate removal in buffer strips (Figure 4.4).

The dynamics of nitrogen and phosphorus retention by soil and vegetation can alter during succession. In newly constructed wetlands, nitrogen retention commences as soon as emergent vegetation becomes established and soil organic matter starts to accumulate: usually within the first 1–3 years. Accumulation of organic carbon in the soil sets the stage for denitrification. After approximately 5–10 years, denitrification removes approximately the same amount of nitrogen as accumulates in organic matter (about 5–10 g m^{-2} yr^{-1} under conditions of low nitrogen loading). Under higher nitrogen loading, the amount of nitrogen stored in accumulating organic matter may double, and nitrogen removal by denitrification may increase by an order of magnitude or more. Accumulation of organic nitrogen and denitrification can therefore provide for reliable long-term removal of nitrogen regardless of nitrogen loading.

Phosphorus removal, on the other hand, tends to be greater during the first 1–3 years of succession when sediment deposition and sorption (absorption and adsorption) and precipitation of phosphorus are greatest. During the early stages of succession, wetlands may retain from 3 g P m^{-2} yr^{-1} under low phosphorus loadings, and as much as 30 g P m^{-2} yr^{-1} under high loadings. However, as sedimentation decreases and sorption sites become saturated, further phosphorus retention relies upon either its accumulation as organic phosphate in plants and their litter, or the precipitation with incoming aqueous and particulate cations such as iron, aluminium and calcium.

Nevertheless, in general, retention of phosphorus tends to be largely regulated by geochemical processes (sorption and precipitation) which operate independently of succession, whereas retention of nitrogen is more likely to be controlled by biological processes (e.g. organic matter accumulation, denitrification) that change in relative significance as succession proceeds.

Surface retention of sediment by vegetated buffer strips is a function of slope length and gradient, vegetation density and flow rates. Construction of effective buffer strips therefore requires detailed knowledge of an area's hydrology and ecology. Overall, restoration of riparian zones in order to improve water quality may have greater economic benefits than allocation of the same land to cultivation of crops.

4.3.4 Wetlands

Wetlands can be used in a similar way to buffer strips as a pollution control mechanism. They often present a relatively cost-effective and practical option for treatment, particularly in environmentally sensitive areas where large waste-water treatment plants are not acceptable. For example, Lake Manzala in Egypt has been suffering from severe pollution problems for several years. This lake is located on the northeastern edge of the Nile Delta, between Damietta and Port Said. Land reclamation projects have reduced the size of the lake from an estimated 1698 km^2 to 770 km^2. The lake is shallow, with an average depth of around 1.3 m.

Five major surface water drains discharge polluted waters into the lake. These waters contain municipal, industrial and agricultural pollutants, which are causing water quality to deteriorate and fish stocks to decline. Recently, efforts have been made to improve water quality in the most polluted of the five drains. This carries

Figure 4.5 A small constructed wetland, planted with common reed (*Phragmites australis*). Such installations are often used to clean water that has undergone primary treatment in a local STW.

wastewater from numerous sources, including sewage effluent from Cairo, waste water from industries, agricultural discharges from farms, and discharges and spills from boat traffic. Several methods for drain water treatment have been proposed, including conventional waste-water treatment plants and other chemical and mechanical methods for aerating the drain water. There are also proposals for construction of a wetland to treat approximately 25 000 m^3 per day of drain water and discharge the treated effluent back to the drain.

The treatment process involves passing the drain water through basins and ponds, designed to have specific retention times. The pumped water first passes through sedimentation basins to allow suspended solids to settle out (primary treatment), followed by a number of wetland ponds (secondary treatment). The ponds are cultivated with different types of aquatic plants, such as emergent macrophytes (e.g. *Phragmites*) with well-developed aerenchyma systems to oxygenate the rhizosphere, allowing the oxidation of ammonium ions to nitrate. Subsequent denitrification removes the nitrogen to the atmosphere.

The waste-water treatment mechanism depends on a wide diversity of highly productive organisms, which produce the biological activity required for treatment. These include decomposers (bacteria and fungi), which break down particulate and dissolved organic material into carbon dioxide and water, and aquatic plants. Some of the latter are able to convey atmospheric oxygen to submerged roots and stems, and some of this oxygen is available to microbial decomposers. Aquatic plants also sequester nitrogen and phosphorus. Species such as common reed (*Phragmites australis*, Figure 4.5) yield a large quantity of biomass, which has a range of commercial uses in the region. Another highly productive species is the water hyacinth (*Eichhornia crassipes*, Figure 4.6). This species is regarded as a serious weed on the lake and is regularly harvested to reduce eutrophication. However, it has a potential role in water treatment due to its high productivity and rapid rates of growth. The resultant biomass could possibly be harvested and used for the production of nutrient-rich animal feed, or for composting and the production of fertilizer. Further research is required to develop practical options.

The passage of water through emergent plants reduces turbidity because the large surface area of stems and leaves acts as a filter for particulate matter. Transmission of light through the water column is improved, enhancing photosynthesis in attached algae. These contribute further to nutrient reduction in through-flowing water. The mixture of floating plants and emergent macrophytes contributes to removal of suspended solids, improved light penetration, increased photosynthesis and the removal of toxic chemicals and heavy metals.

Figure 4.6 Water hyacinth (*Eickhornia crassipes*).

Estimates for the removal of total suspended solids (TSS), biological oxygen demand (BOD), total phosphorus and total nitrogen by the different wetland components are provided in Table 4.2. These suggest that wetlands, combined with sedimentation and ancillary water treatment systems, could play an important part in reducing nutrient loadings.

Table 4.2 Estimated effluent concentrations and removal efficiencies for the Lake Manzala project.

Parameter	Sedimentation pond			Wetland treatment system		
	influent conc./mg l^{-1}	effluent conc./mg l^{-1}	removal efficiency/%	influent conc./mg l^{-1}	effluent conc./mg l^{-1}	removal efficiency/%
TSS	160	32	80	32	6.4	80
BOD	40	24	40	24	18.0	25
total P	5	4	20	4	3.2	20
total N	12	12	0	12	8.4	30
organic N	4	4	0	4	3.8	5
ammonium N	5	5	0	5	3.9	22

4.3.5 Domestic campaigns

An important aspect of efforts to reduce nutrient inputs to water bodies is the modification of domestic behaviour. Public campaigns in Australia have encouraged people to:

- wash vehicles on porous surfaces away from drains or gutters
- reduce use of fertilizers on lawns and gardens
- compost garden and food waste
- use zero- or low-phosphorus detergents
- wash only full loads in washing machines
- collect and bury pet faeces.

These campaigns have combined local lobbying with national strategies to tackle pollution from other sources.

4.4 Reducing nutrient availability

Once nutrients are in an ecosystem, it is usually much harder and more expensive to remove them than tackle the eutrophication at source. The main methods available are:

- precipitation (e.g. treatment with a solution of aluminium or ferrous salt to precipitate phosphates);
- removal of nutrient-enriched sediments, for example by mud pumping; and
- removal of biomass (e.g. harvesting of common reed) and using it for thatching or fuel.

In severe cases of eutrophication, efforts have been made to remove nutrient-enriched sediments from lakes. Lake Trummen in Sweden accumulated thick black sulfurous mud after years of receiving sewage effluent. Even when external loadings of phosphorus were reduced to 3 kg P yr^{-1}, there was still an internal load (i.e. that derived from the lake's own sediment) of 177 kg P yr^{-1}! Drastic action was needed. Eventually nutrient-rich sediment was sucked from the lake and used as fertilizer. The water that was extracted with the sediment was treated with aluminium salts and run back into the lake. This action reduced phosphorus concentrations and improved the clarity and oxygenation of the water. However, removal or sealing of sediments is an expensive measure, and is only a sensible option in severely polluted systems, such as the Norfolk Broads, England.

Figure 4.7 Weed harvesting methods.

Figure 4.8 Mowing and removal of terrestrial vegetation to strip the nutrients contained in the biomass out of the ecosystem.

Removal of fish can allow species of primary consumers, such as the water-flea, *Daphnia*, to recover and control algae. Once water quality has improved, fish can be re-introduced.

Mechanical removal of plants from aquatic systems is a common method for mitigating the effects of eutrophication (Figure 4.7). Efforts may be focused on removal of existing aquatic 'weeds' such as water hyacinth that tend to colonize eutrophic water. Each tonne of wet biomass harvested removes approximately 3 kg N and 0.2 kg P from the system.

Alternatively plants may be introduced deliberately to 'mop up' excess nutrients. Although water hyacinth can be used in water treatment, the water that results from treatment solely with floating macrophytes tends to have low dissolved oxygen. Addition of submerged macrophytes, together with floating or emergent macrophytes, usually gives better results. Submerged plants are not always as efficient as floating ones at assimilating nitrogen and phosphorus due to their slower growth, resulting from poor light transmission through water (particularly if it is turbid) and slow rates of CO_2 diffusion down through the water column. However, many submerged macrophytes have a high capacity to elevate pH and dissolved oxygen, and this improves conditions for other mechanisms of nutrient removal. At higher pH, for example, soluble phosphates can precipitate with calcium, forming insoluble calcium phosphates, so removing soluble phosphates from water. Various species have been used in this way. One submersed macrophyte, *Elodea densa*, has been shown to remove nitrogen and phosphorus from nutrient-enriched water, its efficiency varying according to loading rate. Nitrogen removal rates reached 400 mg N m^{-2} per day during summer, while for phosphorus over 200 mg P m^{-2} per day were removed.

In terrestrial habitats, removal of standing biomass is an important tool in nature conservation. Reduction in the nutrient status of soils is often a prerequisite for re-establishment of semi-natural vegetation, and the removal of harvested vegetation helps to reduce the levels of nutrients returned to the soil (Figure 4.8). However, if the aim is to lower the nutrient status of a nutrient-enriched soil, this can be a very long-term process (Figure 4.9).

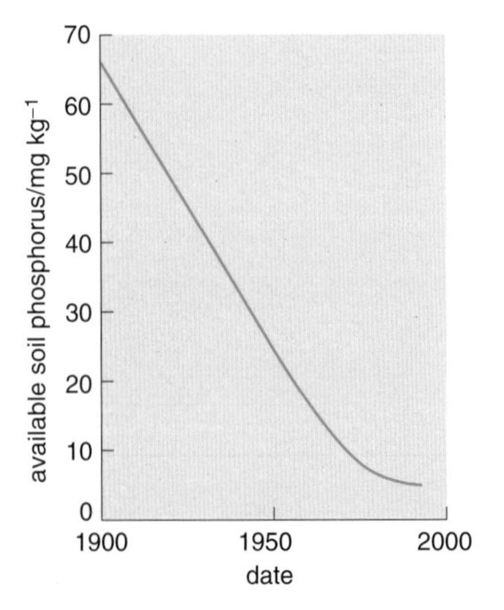

Figure 4.9 Depletion of available phosphorus at Rothamsted long-term experimental husbandry site. Harvesting of arable crops grown without the use of any fertilizer gradually exhausted the reserves of phosphorus in the initially enriched soil.

Question 4.1

A short reach of the River Great Ouse in Bedfordshire was found to contain the following species:

Common name	Scientific name
filamentous alga	*Cladophora* spp.
moss	*Amblystegium riparium*
fool's water-cress	*Apium nodiflorum*
hornwort	*Ceratophyllum demersum*
mare's-tail	*Hippuris vulgaris*
spiked water-milfoil	*Myriophyllum spicatum*
yellow water-lily	*Nuphar lutea*
great water dock	*Rumex hydrolapatham*
water speedwell	*Veronica anagallis-aquatica*
sweet flag	*Acorus calamus*
water plantain	*Alisma plantago-aquatica*
lesser pond sedge	*Carex acutiformis*
reed sweet-grass	*Glyceria maxima*
yellow flag iris	*Iris pseudacorus*
greater duckweed	*Lemna gibba*
broadleaved pondweed	*Potamogeton natans*
bulrush	*Schoenoplectus lacustris*

A similar length of the River Eden in Cumbria was found to have the following species:

Common name	Scientific name
liverwort	*Pellia epiphylla*
moss	*Calliergon cuspidatum*
river moss	*Fontanalis antipyretica*
water horsetail	*Equisetum fluviatile*
white water-lily	*Nymphaea alba*
lesser spearwort	*Ranunculus flammula*
pink water speedwell	*Veronica catenata*
sedge	*Carex acuta*
bottle sedge	*Carex rostrata*
common spike-rush	*Eleocharis palustris*
broadleaved pondweed	*Potamogeton natans*

Using the trophic rank scores in Table 4.1, calculate the mean trophic rank (MTR) for each stretch of river, and comment on whether the watercourse is nutrient-enriched (eutrophic). Assume all the species recorded are of similar abundance and therefore there is no need to weight scores according to relative abundance, as you would do in a real situation.

Question 4.2

List the advantages of preventing eutrophication at source, compared with treating its effects.

5

Summary of Topic 7

1 Eutrophication is a process in which an ecosystem accumulates mineral nutrients. It can occur naturally, but is usually associated with human activity that releases nutrients into the environment.

2 Anthropogenic eutrophication has caused a widespread loss of biodiversity in many systems. Recent attempts to reverse the process are proving difficult and expensive.

3 Symptoms of eutrophication are most readily seen in aquatic systems, where the additional nutrients lead to the explosive growth of algal or bacterial populations. The large biomass produced excludes light from the water and can result in the deoxygenation of the water, killing fish and other animals.

4 In terrestrial systems, additional nutrients boost the productivity of competitive plant species. These then exclude less competitive species by shading them, leading to a decrease in species richness. The humped-back curve describes the relationship between biomass and species richness.

5 Estuaries are particularly prone to eutrophication, and like other aquatic environments can suffer from algal blooms that eliminate other species. Loss of key species, such as seagrass, results in an entire habitat type and all its dependent species disappearing.

6 The main agents of eutrophication are compounds containing the elements phosphorus and nitrogen. It is these elements that, under natural conditions, usually limit the primary production in ecosystems. Increasing their supply therefore increases productivity.

7 Sources of anthropogenic phosphorus entering the environment include sewage discharges, intensive livestock farms and the spreading of artificial fertilizers and animal manures onto agricultural land. The majority of phosphorus comes from point sources.

8 Sources of anthropogenic nitrogen entering the environment include gaseous emissions from car exhausts and power stations and artificial fertilizers applied to agricultural land. The majority of nitrogen comes from diffuse sources.

9 Recent European legislation has tried to limit further eutrophication of the environment by measures such as the stripping of phosphorus from wastewater and the control of nitrogen fertilizer applications in sensitive zones.

10 Living organisms can be used as monitors of the trophic status of ecosystems.

11 Removal of nutrients from an ecosystem in order to reverse the effects of eutrophication is a difficult and expensive undertaking.

Learning outcomes for Topic 7

After you have worked through this topic you should be able to:

1 Describe the principal differences between a eutrophic and an oligotrophic ecosystem. (*Question 1.1*)

2 Explain the mechanisms by which species diversity is reduced as a result of eutrophication. (*Questions 2.1 and 2.2*)

3 Contrast the anthropogenic sources that supply nitrogen and phosphorus to the wider environment, and describe how these sources can be controlled. (*Question 3.1*)

4 Describe how living organisms can be used as monitors of the trophic status of ecosystems. (*Question 4.1*)

5 Compare the advantages and disadvantages of three different methods for combating anthropogenic eutrophication. (*Question 4.2*)

Answers to questions

Question 1.1

(a) Under oligotrophic conditions the stress-tolerant strategy is likely to be the most effective. Stress tolerators have very conservative use of nutrients, retaining them within their tissues and having a slow turnover of tissues to avoid releasing nutrients back to the soil.

(b) In eutrophic conditions the competitive strategy is likely to be the most successful. Competitors are able to increase their growth rate and productivity to make use of extra nutrient availability. They are rather wasteful of the nutrients they acquire due to the high turnover rate of their tissues (both roots and leaves), but they succeed through rapid upward growth which allows them to shade and hence exclude their neighbours.

Question 2.1

River A has a soluble reactive phosphorus concentration in the range 0.1–0.2 mg l^{-1}, which corresponds to the 'degraded' category in Table 2.3. This suggests that the diversity of macrophytes would be less than in the pristine natural state, due to a limited eutrophication effect. Neither form of nitrogen is present at concentrations above the natural range, so primary productivity may become limited by nitrogen rather than phosphorus, limiting the impact of the elevated phosphorus concentration.

River B has a similar concentration of SRP to river A, which would again place it in the 'degraded' category, but a much higher concentration of nitrogen in both its forms, especially nitrate at 12.1 mg l^{-1}, taking it into the 'severe loss of species' category. The elevated availability of both P and N would boost primary production in the watercourse, favouring algal communities and leading to a decline in macrophyte populations and diversity. The more competitive macrophytes may benefit from the increased nutrient availability, but their increased growth would further exclude less competitive species, resulting in lower diversity.

Question 2.2

Seagrass needs sufficient light to photosynthesize effectively. An increase in nutrient levels leads to a greater abundance of phytoplankton in the water column, which increases the turbidity of the water and blocks out the light. As the seagrass beds recede, the exposed sediment may be re-suspended and further increase the turbidity of the water, thereby exacerbating the problem in a classic positive feedback response. Another possible positive feedback loop is the loss of habitat for filter-feeding animals, which lived in the shelter of seagrass beds and helped keep the water clear by consuming microbes.

Question 3.1

Figure 3.1 shows that total phosphorus concentrations had fallen to 0.2 mg l^{-1} in 2000, compared with their peak concentration of 0.36 mg l^{-1} in 1975. In terms of SRP (the form of phosphorus that affects ecosystems most directly), we assume levels have fallen from 0.29 mg l^{-1} to 0.16 mg l^{-1}. Comparing these figures with those in Table 2.3, we see that the SRP concentration put the system in the 'severe loss of species' category in 1975, but only the 'degraded' category in 2000. This suggests that some recovery of macrophyte species would be possible. Actual re-colonization may be a slow process, however. Ecosystems can take many years to come back to equilibrium after a perturbation, and if an algal-dominated state has established, it will inhibit macrophyte recovery.

Question 4.1

The River Great Ouse has a MTR of 3, suggesting it is enriched with nutrients and therefore eutrophic, but it is on the mildest edge of this category so the eutrophication is not severe.

The River Eden has an MTR of 6, indicating that its plant community is composed of species that are moderately sensitive to enrichment, so it can be assumed that this stretch has not undergone substantial eutrophication. The most sensitive species are absent, suggesting that the waters may naturally carry a moderate concentration of nutrients or that some very mild enrichment has occurred.

Question 4.2

Prevention of eutrophication at source compared with treating its effects (or reversing the process) has the following advantages.

- *Technical feasibility*. In some situations prevention at source may be simply engineered by diverting a polluted watercourse away from the sensitive ecosystem, while removal of nutrients from a system by techniques such as mud-pumping is more of a technical challenge.
- *Cost*. Nutrient stripping at source using a precipitant is relatively cheap and simple to implement. Biomass stripping of affected water is labour-intensive and therefore expensive.
- *Habitat availability*. Buffer strips and wetlands can provide a stable wildlife habitat whilst performing a nutrient trapping role on throughflow water. Habitats, whether aquatic or terrestrial, can be compromised in terms of their wildlife value, due to the degree of disturbance involved in biomass stripping.
- *Products*. Constructed wetlands may be managed to provide economic products such as fuel, compost or thatching material more easily than trying to use the biomass stripped from a less managed system.

Acknowledgements for Topic 7 *Eutrophication*

Grateful acknowledgement is made to the following sources for permission to reproduce material in this book:

Figures 1.1, 1.2, 1.3, 1.10, 2.4, 2.8, 2.10, 2.11, 2.17, 3.2, 3.3, 4.2, 4.3, 4.7: Courtesy of the Environment Agency; *Figures 1.4, 1.8a, b, 2.3, 2.12, 2.14, 2.15, 2.16, 3.1, 4.8*: Copyright © Mike Dodd/Open University; *Figure 1.7*: Andy Harmer/Science Photo Library; *Figure 1.8c*: Courtesy of Dr John Madsen/ Minnesota State University; *Figure 1.12*: Copyright © Dr Julian Thompson, Wetland Research Unit, Department of Geography, University College, London; *Figure 2.1*: Copyright © Sukurta Viliaus & Co.; *Figures 2.2, 2.5a, 3.4*: Copyright © Owen Mountford/Centre for Ecology and Hydrology; *Figure 2.5b*: Copyright © Peter Gathercole/Oxford Scientific Films; *Figure 2.6*: Copyright © Alistair MacEwen/Oxford Scientific Films; *Figure 2.7*: Copyright © BBC Worldwide; *Figure 2.18*: With permission of Professor Chris Kjeldsen, Biology Department, Sonoma State University; *Figure 2.24*: Graham Day; *Figure 4.4*: Copyright © Peter Leeds-Harrison/Cranfield University; *Figure 4.5*: Copyright © ARM Limited. Reproduced with permission of ARM Limited, Rydal House, Colton Road, Rugeley, Staffordshire, WS15 3HF, UK, www.armreedbeds.co.uk, designers of reed beds and wetlands for wastewater treatment; *Figure 4.6*: Pisces Conservation Limited.

Every effort has been made to trace all the copyright owners, but if any has been inadvertently overlooked, the publishers will be pleased to make the necessary arrangements at the first opportunity.

TOPIC 8

ACID RAIN

Jack Cosby, Tim Allott, Nancy Dise, and Lesley Smart

Introduction

1

Acid rain has been one of the more prominent public environmental issues since the 1970s, and continues to be of major interest in the public, government and scientific arenas. The causes and effects of acid rain have been debated in both scientific and political circles, and the problem is one of international significance. Concerns over the possible effects of acid rain were first voiced in Scandinavia in the early 1970s. These concerns grew from observations of declines in the fisheries of southern Scandinavia. Records going back almost a century are available of the catches of Atlantic salmon from rivers in southern Norway. From about 1910, these records show an alarming decline in the amount of salmon caught (Figure 1.1). This decline is not apparent in the amount of salmon caught in the rivers of western and northern Norway. Brown trout also began to disappear from lakes in the mountains of southern Scandinavia, and by 1975 almost 50% of the lakes in this area were believed to have lost their brown trout populations (Figure 1.2).

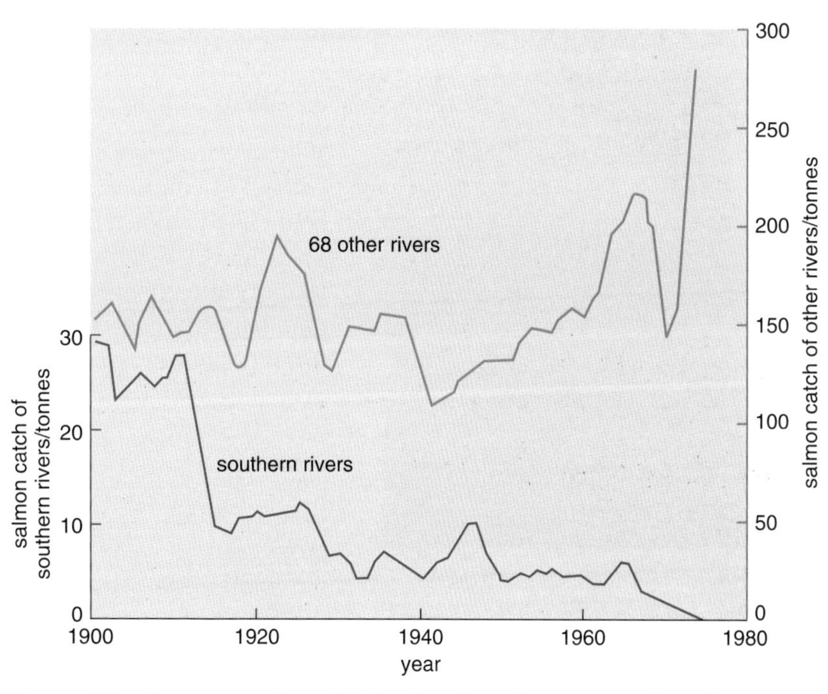

Figure 1.1 Average catches of Atlantic salmon from seven rivers in southern Norway receiving acid deposition, and 68 other rivers that did not receive acid deposition, from 1900 to 1975. (1 tonne = 10^3 kg)

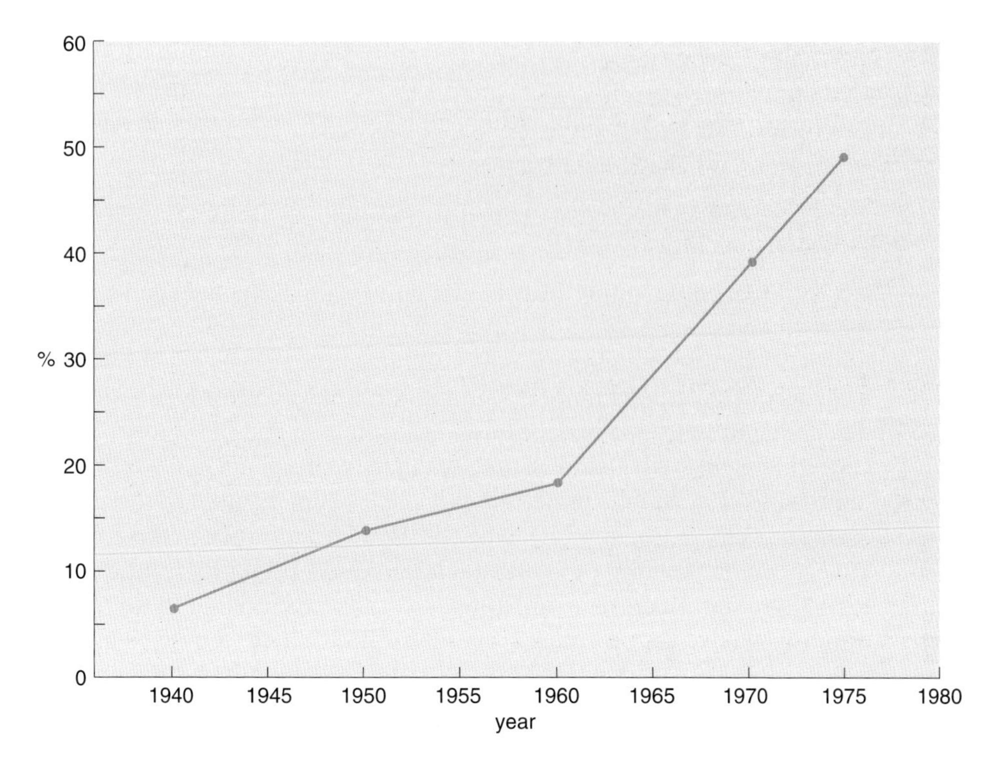

Figure 1.2 The estimated percentage of lakes in southern Scandinavia that have lost their population of brown trout, from 1940 to 1975.

Figure 1.3 Black smoke from a power plant spews sulfur dioxide into the air over the Bosnian city of Tuzla, 25 December 1995.

Figure 1.4 Eroded stonework, Wells Cathedral, Somerset, England.

Norwegian scientists noticed that, when compared to similar lakes in northern Scandinavia, the lakes that had lost their fish stocks were unusually acidic, often having a pH of less than 5.0. The weather in southern Norway is derived from air masses passing over central and Western Europe. The scientists argued that the lakes had acidified following the effects of acid rainfall originating from the industrial areas of central and western Europe, and that this acidification had resulted in the deaths of fish such as brown trout and Atlantic salmon.

When these arguments were first made, they sparked off a considerable amount of controversy, particularly in countries that were accused of being major contributors to the acid rain falling over southern Scandinavia, notably the UK. Some scientists and politicians argued that lake acidification and reduction in fish catches were not the result of acid rain, but were caused by other processes such as afforestation (growing forests), heather burning, a change in land use, or even a natural increase in acidification over time.

During the 1980s, a huge research effort was mounted to study the causes and effects of acid rain. The UK Central Electricity Generating Board (CEGB) and British Coal financed a five-year research programme, which was jointly run by the UK and Sweden. This research programme, known as the Surface Waters Acidification Programme (SWAP), made a considerable contribution to the understanding of acid rain and its subsequent effects. A summary of the main conclusions of the Programme is contained in Appendix 1.

We now know that acid rain, more properly called acid deposition, is mainly caused by anthropogenic emissions of air pollutants from industrial sources such as power stations (Figure 1.3). We also know that acid deposition has been responsible for significant damage to the environment. The effects of acid deposition include:

- the acidification of soils, rivers and lakes;
- the loss of fisheries in freshwaters;
- the decline in the health of natural vegetation and forests;
- damage to historic buildings (Figure 1.4).

Up until the 1990s, sulfur dioxide was universally acknowledged to be the pollutant principally responsible for acid deposition. Since then, however, the roles of nitrogen oxides and reduced nitrogen in acidification have become relatively more important in Europe and North America, as the amount of nitrogen in deposition has increased and the amount of sulfur has decreased. Oxidized nitrogen in the atmosphere in the form of nitrogen oxides, NO_x, is produced by industrial combustion of fossil fuels and vehicle (mainly car) exhausts. Reduced nitrogen in the atmosphere in the form of ammonia, NH_3, results primarily from agricultural fertilizers and animal wastes. Like SO_2, both of these forms of nitrogen have the potential to cause acidification of soils and surface water.

It is much more complicated to evaluate the effects of enhanced nitrogen deposition on the biosphere than the effects of sulfur deposition.

❍ Apart from acidification, can you think of another way in which nitrogen deposition could potentially affect terrestrial and aquatic ecosystems?

● Since the productivity of many terrestrial and aquatic ecosystems is limited by nitrogen, additional nitrogen deposition could act like a fertilizer.

As long as this oxidized and reduced nitrogen is utilized by plants when it is deposited from the atmosphere, then there is little or no acidification of surface waters due to the nitrogen (although there may be some acidification of soils due to plants exchanging H^+ for NH_4^+). However, scientists are concerned that, as a result of continued nitrogen deposition, the terrestrial ecosystem may become saturated with nitrogen and uptake by plants will cease to match the deposition. If this happens, and it already has happened in some places, the deposited nitrogen will contribute to the soil and surface water acidification already produced by sulfur deposition.

In this topic we will concentrate on the role played by sulfur in acid deposition. This is because the mechanisms and effects of acidification by sulfur deposition are broadly similar to those of nitrogen deposition, without major confounding biological effects. We shall examine the main chemical processes that lead to the formation of sulfuric acid in atmospheric deposition and the resulting acidification of soils and surface waters. We shall also review the research that finally proved the connection between acid deposition and surface water acidification. Finally, we shall investigate the mathematical models that are used to predict the effects of increases and decreases in acid deposition on soils and waters.

1.1 Summary of Section 1

1 In the 1970s it was observed that lakes in southern Scandinavia showed a major decline in fish stocks. This decline can be correlated with an increase in lake acidity.

2 The acidification of surface waters is due to acid deposition caused by anthropogenic emissions of gaseous pollutants, particularly sulfur dioxide, SO_2.

3 As sulfur deposition decreases, nitrogen oxides, NO_x, and ammonia, NH_3, contribute an increasing proportion to acid deposition in the UK, Europe and North America. However, the NO_3^- and NH_4^+ ions that are subsequently formed can largely be taken up harmlessly by the roots of plants, and so the environmental impact in many areas has not yet been great.

2

Sources of sulfur

2.1 Sulfur in the environment

Sulfur occurs naturally in almost all parts of the environment. It is present within the Earth's crust, the oceans, the biosphere and the atmosphere:

- The average concentration of sulfur in the Earth's crust is 260 ppm; it is most commonly found in rocks within volcanic regions, often as sulfides of metals.
- Sulfur is an important element in seas and oceans, usually in the form of neutral salts. The average concentration of sulfur in seawater is 905 ppm.
- Sulfur is an important minor constituent of amino acids and proteins in living organisms, constituting 0.07% (700 ppm) of the biosphere in terms of mass.
- In the atmosphere, sulfur is most commonly found as sulfur dioxide, derived from both natural and anthropogenic sources. Levels of SO_2 in unpolluted air are as low as 1 part per billion by volume (ppbv), whereas in polluted urban areas SO_2 concentrations can reach 1–2 ppm by volume (ppmv). Since the implementation of significant SO_2 control measures, the ambient SO_2 concentration in urban areas in Europe and North America has been reduced significantly. Overall, the average concentration of sulfur in the global atmosphere is about 0.6 ppmv.

Sulfur is used to make sulfuric acid, vulcanized rubber, dyes and various chemicals. Sulfur compounds are used in many medicines.

2.2 Sulfur in the atmosphere

Although the average concentration of sulfur in the atmosphere is known, the precise quantity and the fractions of the total derived from different sources are uncertain. The sources of sulfur can be split into two types — natural and anthropogenic sources. Globally, natural and anthropogenic sulfur emissions are of the same order of magnitude. A schematic representation of the global sulfur cycle is shown in Figure 2.1.

2.2.1 Natural sources of atmospheric sulfur

There are three main natural sources of atmospheric sulfur:

- the oceans;
- volcanic emissions;
- the decay of organic matter.

The oceans are by far the most important natural source of atmospheric sulfur. Sulfur is primarily present within oceans in the form of dissolved salts, principally sulfates. Sulfate is released into the atmosphere as sea spray. This evaporates, leaving fine particles containing sulfur in the atmosphere. The gaseous forms of sulfur enter the atmosphere through gaseous exchange at the ocean's surface.

In nutrient-rich areas of the oceans, a further source of sulfur is the production of dimethyl sulfide (DMS), $(CH_3)_2S$, by decomposing marine phytoplankton and zooplankton (floating microscopic plants and animals). Once released into the

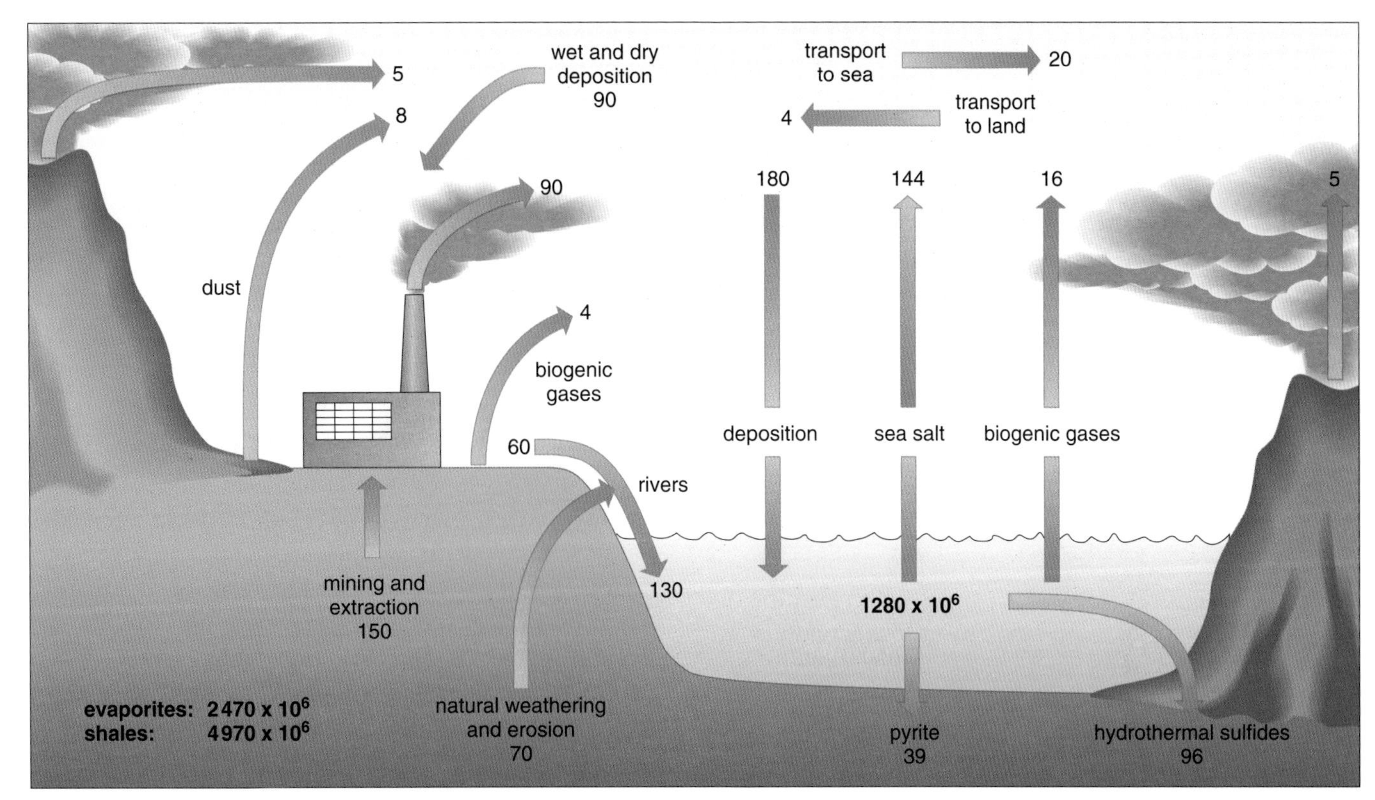

Figure 2.1 The present-day global sulfur cycle. Nearly all the sea salt, volcanic gases, and biogenic gases released from the sea are re-deposited in the sea. All reservoirs (bold values) are expressed in units of 10^9 kg S and annual fluxes in units of 10^9 kg S yr^{-1}.

atmosphere, DMS is oxidized to form methylsulfonic acid, CH_3SO_3H, and SO_2. DMS is responsible for the sometimes pungent 'sea air' smell of beaches and shores. DMS is the major biogenic gas emitted from the sea, with a flux of around 15×10^9 kg S per year.

The oceans also release small amounts of hydrogen sulfide gas, H_2S, although the major source of H_2S is on land. The formation of H_2S is a characteristic feature of the decomposition of organic matter in anaerobic marine sediments. In ecosystems such as saltmarshes, mangrove swamps or shallow, stagnant estuaries, the H_2S produced can escape oxidation and be released to the atmosphere. Decomposition of organic matter (represented in Equation 2.1 by the ethanoate ion) in the presence of sulfates produces carbon dioxide and hydrogen sulfide gases dissolved in the seawater in contact with the sediments:

$$\underset{\text{ethanoate}}{CH_3COO^-(aq)} + SO_4^{2-}(aq) + 3H^+(aq) = 2CO_2(aq) + 2H_2O(l) + H_2S(aq) \qquad (2.1)$$

Hydrogen sulfide released into the atmosphere from these sediments is rapidly oxidized to form SO_2 gas.

- What happens to the hydrogen ions on the left-hand side of Equation 2.1?
- They are neutralized (they don't appear on the right-hand side of the equation).

Anaerobic oxidation processes, such as sulfate reduction and denitrification, consume hydrogen ions, and can thus reduce acidity (or increase alkalinity) in surface waters.

Volcanic activity can also result in significant emissions of SO_2 and H_2S into the atmosphere. During volcanic eruptions, material can be injected high into the atmosphere and can be transported rapidly over vast distances. Volcanic emissions, however, are somewhat sporadic, and total global sulfur emissions from this source are relatively low.

Sources of natural emissions of sulfur are widely dispersed over the planet. It has therefore proved difficult to establish accurately the rates of production from these sources, and all of the above estimates have fairly large uncertainties.

2.2.2 Anthropogenic sources of atmospheric sulfur

Sulfur emissions are also caused by human activity. These emissions originate from two principal sources:

- First, and most importantly, they derive from the burning of fossil fuels for heat and electricity generation. The fossil fuels of relevance are coal and oil.
- The second anthropogenic source is through industrial processes such as the smelting of iron and other metals whose ores contain significant amounts of sulfur, and through the manufacture of sulfuric acid.

❍ All fossil fuels contain sulfur from at least one common source. What is it?

● Sulfur is one of the macronutrients necessary for all life. It is incorporated into the amino acids of all living things, including the vascular plants that eventually form coal and the algae and other plankton that form oil.

The burning of fossil fuels is clearly responsible for the major component of anthropogenic SO_2 emissions. Coal and oil were formed by the decomposition of plants and marine plankton, respectively, during the Carboniferous Period (about 300 million years ago). On death, these organisms were transformed over millions of years by the processes of decay and compaction into deposits of fossil fuel. Fossil fuels retain some of the compounds that were originally present in the bodies of the living organisms. When fossil fuels are burnt, the carbon and sulfur from these compounds are released into the atmosphere as CO_2 and SO_2.

In coal, sulfur is present in both inorganic and organic forms. The total sulfur content of coal can be as high as 6–8%. However, 40–60% of this is in the form of inorganic iron pyrites, FeS_2, which can be removed easily by physical means — pulverizing and washing coal before it is burnt in power stations reduces the sulfur content to less than 2%. In general, oil has a lower sulfur content than coal. In crude oil, the sulfur content varies from trace levels to several per cent; in fuel oils the typical content is 0.7%.

As with natural emissions, the industrial release of sulfur into the atmosphere is difficult to quantify accurately, mainly due to variations from year to year and changing patterns through time. Globally, total industrial emissions are somewhat lower than those from natural sources (Figure 2.1), although in the Northern Hemisphere they exceed natural emissions. An important feature of anthropogenic emissions is that two-thirds of them originate from the urban and industrial areas of Europe and North America. This leads to significant elevations of SO_2 concentrations above natural background levels in these areas. The consequences of this air pollution, and the associated environmental effects, are discussed in Section 5.

2.3 Trends in anthropogenic sulfur emissions

Anthropogenic emissions of SO_2 have increased dramatically since the Industrial Revolution, when the burning of fossil fuels first became significant on a global scale. Figure 2.2a shows the Asian, European (including the UK), North American and global anthropogenic emissions of SO_2 from 1850 to 1990. The graph shows an almost exponential increase in global SO_2 emissions up until the 1980s. Of particular note is the acceleration in the amount of SO_2 emissions after the Second World War. This is due to the post-war increase in demand for electricity in the West, which resulted in an increase in the quantities of fossil fuels burnt.

In 1980, global emissions of SO_2 were over 20 times larger than those in 1860. Since 1980, however, emissions in countries within Europe and North America have declined as a result of legislation that imposed limits on sulfur emissions (discussed in Section 5.3.2). Restrictions on these emissions were a direct result of the scientific research that established the linkages between sulfur emissions, acid

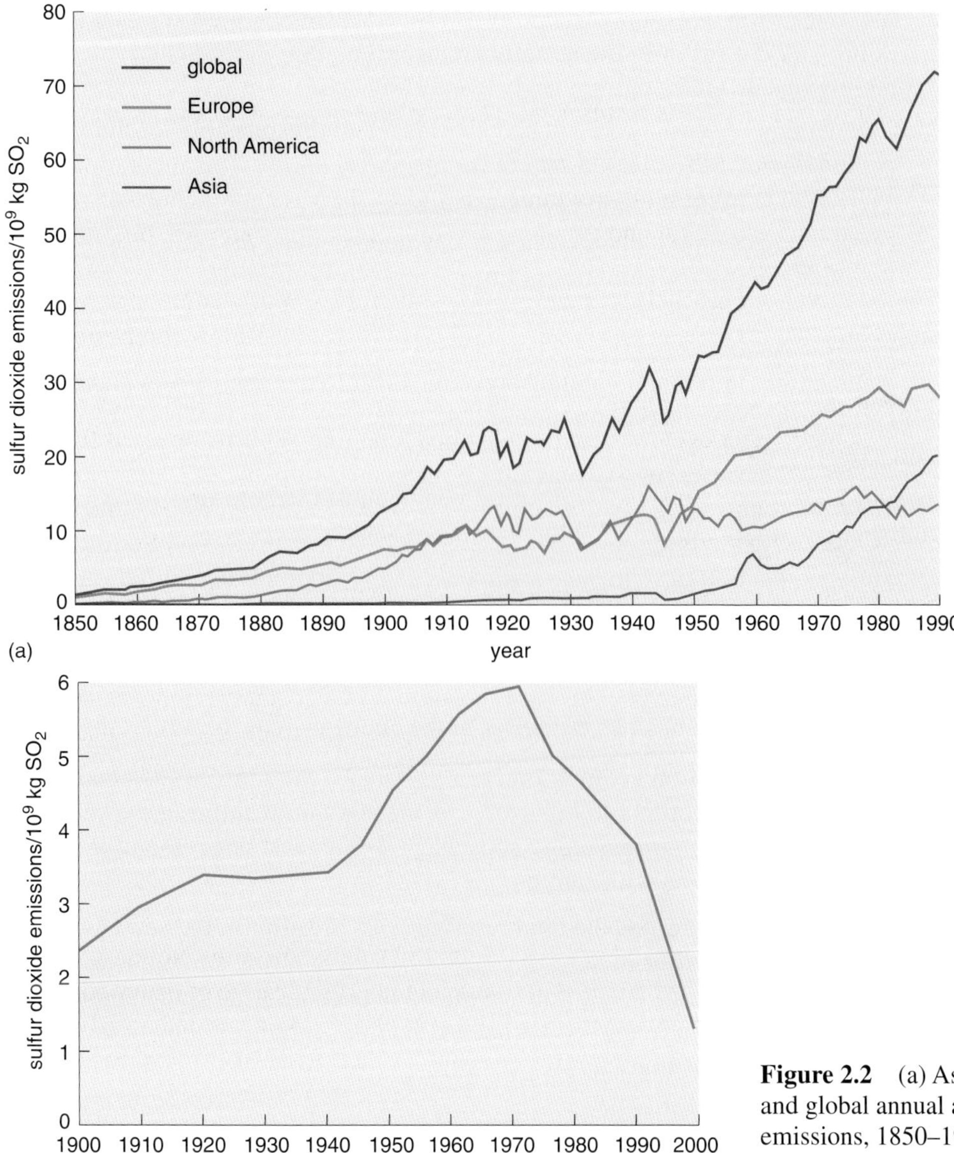

Figure 2.2 (a) Asian, European, North American and global annual anthropogenic sulfur dioxide emissions, 1850–1990; (b) UK annual anthropogenic sulfur dioxide emissions, 1900–1999.

deposition, and the acidification of soils and waters. As a reflection of this, global SO_2 emissions also declined for a few years in the early 1980s. However, global emissions have since begun to increase steeply again, reflecting a new major global source of anthropogenic SO_2 emissions: Asia.

Global and international compilations require a high level of data gathering and co-ordination, so those estimates tend to be older than emissions data from single countries. Figure 2.2b shows SO_2 emissions for the UK extending to 1999. The graph shows a steep decline in SO_2 emissions, starting in the early 1970s. SO_2 emissions in 1999 were lower than at any time in the previous 100 years.

2.4 Summary of Section 2

1 Sulfur occurs naturally in all parts of the environment, including the atmosphere, oceans, soils, rocks and living organisms. Natural sources of sulfur include sulfides of metals in rocks, sea spray, biogenic gases, and volcanic activity.

2 Almost all the sulfur in the atmosphere occurs as SO_2. Natural and anthropogenic emissions are fairly evenly balanced.

3 The burning of fossil fuels (oil and coal) is mainly responsible for the anthropogenic SO_2 released into the atmosphere.

4 The changes in the amount of SO_2 in the atmosphere over time during the 20th century reflect both human economic development and the influence of pollution control technologies in Europe and North America.

Question 2.1

(a) A minor, but important, natural source of sulfur in the atmosphere is the biological production of H_2S through sulfate reduction. Identify where the major sulfur flux from this process is shown on Figure 2.1.

(b) In which ecosystems on Earth would you be most likely to find significant biological production of H_2S? Why?

(c) Sulfate reduction in some lake sediments can be an important buffer on acidity due to acid rain. Explain why that is the case.

Question 2.2

One of the major natural sinks for sulfur in the atmosphere is the sea.

(a) From the information given in Figure 2.1, balance the sources and sinks of sulfur for the sea. According to Figure 2.1, is the amount of sulfur in the sea increasing, decreasing, or remaining constant? Give at least one reason for viewing your conclusion with caution.

(b) According to Figure 2.1, the total reservoir size of sulfur in the sea (overwhelmingly as sulfur in SO_4^{2-}) is 1.28×10^{18} kg S. Using a rough average of the total input and output fluxes you calculated in part (a), what is the approximate residence time of sulfur in the sea? What are the implications of this for the long-term fate of acid deposition?

Sulfur emissions and acid deposition

3

In Section 2 we looked at the mechanisms by which sulfur is emitted to the atmosphere and stressed the importance of anthropogenic sources of sulfur, particularly the burning of fossil fuels in industrial and urban regions of the world. During fossil fuel combustion, sulfur combines with the oxygen present in the air to produce sulfur dioxide, SO_2. Once emitted, the gaseous pollutants such as SO_2 are transported by wind away from the emission source, such as a power station or industrial plant. The emission plume spreads out into an expanding cone, which travels with the prevailing wind (Figure 3.1). It is usually confined within the well-mixed polluted layer of the atmosphere, which can vary from a depth of a few hundred metres to about 2 km, depending on conditions. The plume gradually loses its shape as it travels further from the source and is distorted by atmospheric turbulence and wind shear. The SO_2 emissions are therefore gradually diluted in the atmosphere and mixed with air.

A plume from a chimney 200 m high will typically make first contact with the ground some 10–20 km downwind, depending on the wind speed, where some of the acid pollutants will be deposited on the ground in a dry form (gaseous or particulate). By this time the SO_2 concentrations will be about 10 000 times less than at the point of emission due to dilution of the plume. The remaining pollutants may travel for a few days, covering hundreds, or even thousands, of kilometres, during which time a series of chemical processes take place that transform the SO_2 into secondary pollutants. These transformations involve reactions with various oxidants to form sulfuric acid, and can take place in both wet and dry air.

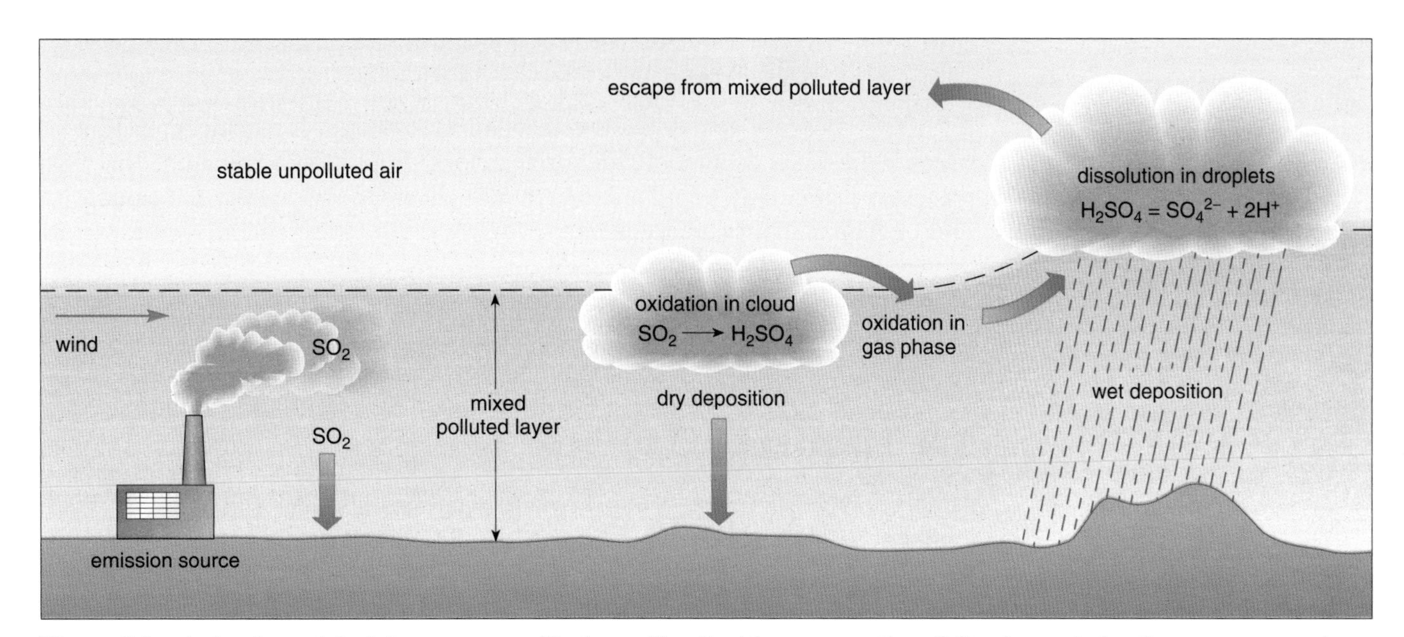

Figure 3.1 A simple model of the processes affecting sulfur dioxide concentrations following emission from a power station chimney.

You will come across the terms *acid precipitation* (which includes rain and snowfall) and *acid deposition* used in the context of acid rain. **Acid deposition** includes all acids produced in the atmosphere, which, in one form or another (rain, snow, hail, mist, etc.), are deposited on the Earth's surface. There are two principal forms of acid deposition — dry deposition and wet deposition.

Dry deposition

This is the direct deposition onto the ground and vegetation of untransformed SO_2 gas, by absorption and adsorption, and of SO_2 that has been transformed into sulfate particles. Dry-deposition rates are calculated from the product of the SO_2 concentration in the air and the characteristic deposition velocity of the molecule or particle. Typically the amount of SO_2 in a boundary layer (mixed layer) extending about 1 km above the surface of the Earth will be depleted by dry deposition at a rate of about 3% per hour, but the depletion rate may decrease with increasing distance from the source. At any distance from the source, the amount of SO_2 remaining to be deposited depends on wind speed, on whether the plume has encountered rain clouds, and on the rate at which the SO_2 is oxidized to sulfate during its travels. In the UK, the pattern of dry deposition closely follows that of SO_2 concentrations. The higher amounts of dry deposition are found close to emission sources such as power stations and industrial areas. In the UK, the largest value of dry deposition of more than $3\ g\ S\ m^{-2}\ yr^{-1}$ occurs in the industrial Midlands of England. In areas remote from emission sources, such as northern Scotland, dry deposition may be of relatively little importance.

Wet deposition

This involves the deposition of transformed oxides of sulfur in aqueous form (as sulfuric acid), either as rainfall or snowfall. Wet deposition removes sulfur from the atmosphere much faster than dry deposition — moderate rain can remove more material in an hour than dry deposition removes over 2–3 days. Wet deposition is particularly important in areas of high total rainfall such as the Lake District, the Highlands of Scotland, the Appalachians in the eastern USA, and other mountainous areas.

A further process of acid deposition that can be important is 'occult deposition', which is the direct capture of wind-driven cloud droplets by vegetation. This process is particularly important in high-altitude areas, where mist and cloud caps on mountain tops are frequent.

Sections 3.2 and 3.3 consider the chemical processes taking place in the atmosphere and how these processes lead to acid deposition. Before considering polluted rainfall, however, it is useful to consider the acidity of natural, unpolluted rain.

3.1 Natural rainwater acidity

Acidity is measured in units of pH, where every increase of 1 unit in pH represents a tenfold *decrease* in the concentration of H^+ ions. The pH of pure water is 7.0. Water within the atmosphere, however, comes into contact with many chemical species, and is therefore not pure. Indeed, clean or 'pristine' rainwater, unaffected by pollution from anthropogenic sources, is naturally acidic. Carbon dioxide, CO_2, is the major 'natural' gas in the atmosphere that contributes to acidity. Once dissolved, CO_2 reacts with water to produce hydrogen ions, H^+, and hydrogen carbonate ions, HCO_3^-:

$$CO_2(g) = CO_2(aq) \quad (3.1)$$

$$CO_2(aq) + H_2O(l) = H^+(aq) + HCO_3^-(aq) \quad (3.2)$$

Hydrogen carbonate ions can dissociate further to produce carbonate ions:

$$HCO_3^-(aq) = H^+(aq) + CO_3^{2-}(aq) \quad (3.3)$$

Reaction 3.2 has by far the most effect on the natural acidity of rainwater. It is apparent from these reactions that the pH of water within the atmosphere will be lower than that of pure water as a result of the dissolution of CO_2. Unpolluted rainwater in equilibrium with atmospheric CO_2 has a pH of 5.65.

- ❍ About how much more acidic is rainwater with a pH of between 5 and 6, in comparison to pure water?
- ● Since there is a tenfold increase in acidity with every unit decrease in pH, unpolluted rainwater (pH between 5 and 6) is between 10 and 100 times more acidic than pure water (pH 7).

However, this value varies naturally due to the effects of other chemical species in the atmosphere.

In areas where calcareous dust is blown into the atmosphere, such as the North American prairies, rainwater pH can be significantly higher than 5.6. Conversely, naturally occurring gaseous sulfur and nitrogen oxides, such as those from volcanoes or biological activity, can depress the pH of rainwater. As a result, rainfall pH in areas remote from pollution sources may be as low as pH 5.0 (or even lower, if directly influenced by volcanic plumes), or higher than pH 7.0.

3.2 Wet-phase transformation of sulfur dioxide

Sulfur dioxide will dissolve in, and react with, rainwater in a manner analogous to carbon dioxide:

$$SO_2(g) = SO_2(aq) \quad (3.4)$$

$$SO_2(aq) + H_2O(l) = H^+(aq) + HSO_3^-(aq) \quad (3.5)$$

$$HSO_3^-(aq) = H^+(aq) + SO_3^{2-}(aq) \quad (3.6)$$

The solubility of SO_2 in water is much lower than the solubility of CO_2 in water; only about 1% of atmospheric SO_2 will dissolve in rainwater. In polluted areas, however, the sulfur content of rainwater is far greater and the pH is far lower than can be accounted for simply by the dissolution of SO_2 in water. The solution to this apparent paradox is the presence in the atmosphere of powerful oxidants such as ozone and hydrogen peroxide, H_2O_2. In the atmosphere, an important source of H_2O_2 is the recombination of hydroperoxy radicals, $HO_2•$, which can be represented as:

$$HO_2• + HO_2• \longrightarrow H_2O_2 + O_2 \quad (3.7)$$

❍ Thinking back to your study of Topic 2 *Atmospheric Chemistry and Pollution*, what is the main source of HO_2• radicals?

● HO_2• (and other peroxy radicals) are formed during the atmospheric oxidation of carbon monoxide, methane and other non-methane hydrocarbons.

When SO_2 comes into contact with cloud or raindrops, the SO_2 dissolves in the water droplets to form the hydrogen sulfite ion, HSO_3^-, and sulfite ion, SO_3^{2-} (Equations 3.5 and 3.6, respectively). Hydrogen sulfite is then oxidized to form the hydrogen sulfate anion, HSO_4^- (Equations 3.8 and 3.9). The aqueous oxidation of HSO_3^- to HSO_4^- can be achieved by either ozone, O_3, or hydrogen peroxide, H_2O_2, dissolved from the gas phase. The mechanisms involved are complex, but the net effect can be represented by the following overall reactions:

$$HSO_3^-(aq) + O_3(aq) = HSO_4^-(aq) + O_2(g) \quad (3.8)$$

$$HSO_3^-(aq) + H_2O_2(aq) = HSO_4^-(aq) + H_2O(l) \quad (3.9)$$

HSO_4^- can further dissociate to the sulfate ion, SO_4^{2-}:

$$HSO_4^-(aq) = H^+(aq) + SO_4^{2-}(aq) \quad (3.10)$$

Hydrogen ion production in the processes leading to Equation 3.10 (Equations 3.5 and 3.6) brings about a large increase in the acidity of cloud and rainwater. The conversion rate of SO_2 to SO_4^{2-} in the wet-phase reaction with ozone is very pH-dependent and, although fast at pH 5, the reaction falls off rapidly with increasing acidity. Oxidation by H_2O_2 then becomes the more important mechanism for producing SO_4^{2-} in most polluted situations. The conversion rate of SO_2 to SO_4^{2-} by H_2O_2 is independent of pH, and is extremely rapid. An important constraint is the supply of H_2O_2. In very SO_2-polluted air, H_2O_2 occurs in low concentrations relative to SO_2, such that the availability of H_2O_2 becomes the limiting control on acidity production under a wide variety of climatic conditions.

3.3 Dry-phase transformation of sulfur dioxide

As well as wet-phase transformation, sulfate ions can also be formed by dry processes. The rate of conversion of SO_2 by dry processes is much slower than by wet processes, but dry processes are nonetheless very important overall because they take place all the time, and are not limited to wet weather conditions. The production of sulfuric acid in a dry atmosphere can take place by three different gas-phase pathways, two of which are dependent on the presence in the atmosphere of hydroxyl radicals, HO•, or electronically excited atomic oxygen atoms, O* (see Box 3.1).

Box 3.1 Production of hydroxyl radicals

As you saw in Topic 2 *Atmospheric Chemistry and Pollution*, an atom of electronically excited oxygen, O*, is produced when ozone absorbs ultraviolet radiation of wavelengths less than 320 nm:

$$O_3 + hf \longrightarrow O_2 + O^*$$

Most of the O* atoms generated in this way lose energy (and return to the ground state) when they bump into other molecules in the air. However, the excess energy possessed by an excited oxygen atom has one very important consequence: if an O* atom happens to collide with a water molecule, it has enough energy — *unlike a ground-state oxygen atom* — to strip off a hydrogen atom from the water molecule:

$$O^* + H_2O \longrightarrow HO\bullet + HO\bullet$$

The hydroxyl radical, HO•, is thus formed by the reaction of an excited oxygen atom, O*, with a water molecule. In forming the HO• radical, one hydrogen atom is stripped away *complete with its single electron*. Neither fragment carries an electric charge. By contrast, in the more familiar dissociation reaction:

$$H_2O(l) = H^+(aq) + OH^-(aq)$$

one hydrogen departs as a bare proton, H^+, its electron being retained by the hydroxide ion, OH^-.

The first, and by far the most important oxidation pathway involves a sequence of reactions initiated by hydroxyl radical attack on SO_2; the first step produces a hydrogen sulfite radical, $HSO_3\bullet$:

$$HO\bullet + SO_2 \longrightarrow HSO_3\bullet \qquad (3.11)$$

The hydrogen sulfite radical then reacts with a molecule of oxygen to produce SO_3 and a hydroperoxyl radical, $HO_2\bullet$:

$$HSO_3\bullet + O_2 \longrightarrow SO_3 + HO_2\bullet \qquad (3.12)$$

The subsequent reaction of SO_3 with water vapour leads eventually to the production of sulfuric acid, as shown in the following overall reaction:

$$SO_3(g) + H_2O(g) = H_2SO_4(l) \qquad (3.13)$$

The second of the three routes to sulfuric acid production involves the formation of sulfur trioxide through the reaction of SO_2 with electronically excited atomic oxygen, O*:

$$O^* + SO_2 \longrightarrow SO_3 \qquad (3.14)$$

The sulfur trioxide thus formed again reacts with water vapour to produce sulfuric acid (Equation 3.13). This reaction pathway depends upon the highly energetic nature of the excited oxygen atom, O*. Oxygen in the ground state does *not* react with the SO_2 molecule.

The third process of dry-phase transformation depends on the presence of hydrocarbon radicals. As you saw in *Atmospheric Chemistry and Pollution*, hydrocarbons are present in the air as pollutants, and they react with HO• radicals to form hydrocarbon radicals that can, in turn, oxidize SO_2 to SO_3. Once again, the sulfur trioxide is converted to sulfuric acid by reaction with water vapour (Equation 3.13). This route to sulfuric acid production is generally of limited significance except in heavily polluted urban and industrial areas.

The conversion rates of SO_2 to H_2SO_4 for dry-phase reactions are significantly lower than those for the wet-phase reactions discussed above. However, clouds and rain are not present all of the time, so that, overall, dry deposition can be as important — or even more important — than wet-phase deposition. In the presence of water (e.g. cloud or raindrops), sulfuric acid formed by dry-phase reactions will dissolve and dissociate, yielding H^+ and HSO_4^- ions:

$$H_2SO_4(aq) = H^+(aq) + HSO_4^-(aq) \qquad (3.15)$$

HSO_4^- can dissociate further to SO_4^{2-} and H^+ ions (see Equation 3.10).

3.4 Summary of Section 3

1 Natural rainwater is not neutral but has a pH of about 5.65 due to dissolved CO_2.

2 SO_2 has a low solubility in water. The production of hydrogen ions and sulfate ions in water from SO_2 is greatly enhanced by the presence of oxidants like ozone and hydrogen peroxide in polluted air.

3 Sulfuric acid formed by the transformation of SO_2 is deposited on the ground by various wet and dry processes.

4 In dry conditions, SO_2 is oxidized by excited atomic oxygen, O*, and by HO• radicals to form sulfur trioxide, SO_3; this can then react with water vapour to form sulfuric acid.

Question 3.1

The acidity of deposition and the amount of acid falling on ecosystems are often enhanced in mountainous areas. Why do you think this is the case?

Acid deposition and acidification

4

We have seen how anthropogenic sulfur emissions can lead to increases in both the acidity of precipitation and the deposition of sulfuric acid from the atmosphere. We now consider the mechanisms by which these deposited acids can lead to the acidification of streams, rivers and lakes. When acid deposition falls on the land surface of the Earth, some of the acidity is deposited directly into the surface waters, but most of the acid deposition falls on the land itself. This acid deposition is washed into the soil, and its subsequent chemistry is greatly altered by interactions that take place within the soil. The sensitivity of a freshwater system to acidification thus largely depends on the chemical properties of the soil. We therefore need to look at the important processes of acidification that take place in the soil, before considering freshwater acidification.

4.1 Soil acidification

The processes that occur within soils that lead to soil and freshwater acidification are highly complex. They involve:

- the exchange of cations between water and soil;
- the uptake and leaching of nutrients by plants;
- chemical transformations by micro-organisms;
- inorganic reactions, such as those involving aluminium.

The degree of acidification of a soil and the soil water depends on the acidity of the rainwater deposition, the chemical properties of the soil, and the water flowpaths within the soil (that determine the time available for chemical and biological processes to operate).

The deposition inputs that we are concerned with here are H^+, HSO_3^-, HSO_4^- and SO_4^{2-} ions, and we need to consider the principal processes in soils involving these ions.

4.1.1 Soil structure

As you already know, soil usually consists of several horizons that have been formed by the actions of weathering, the deposition and decomposition of various materials (such as leaves, and rainfall containing dissolved solids), and the formation and movement of compounds from one layer to another by percolating soil water. Figure 4.1 shows the profile of a spodosol (podzol). Spodosols are the typical leached, acidic soils found in areas most affected by acid rain, such as forests growing in areas dominated by granite bedrock.

In spodosols the A horizon usually has a pale ashy layer underneath it, designated the E horizon. The E horizon is a leached mineral layer of predominantly sand or silt with little organic matter; most of the clay, aluminium and iron have been leached out to lower layers, forming soluble complexes

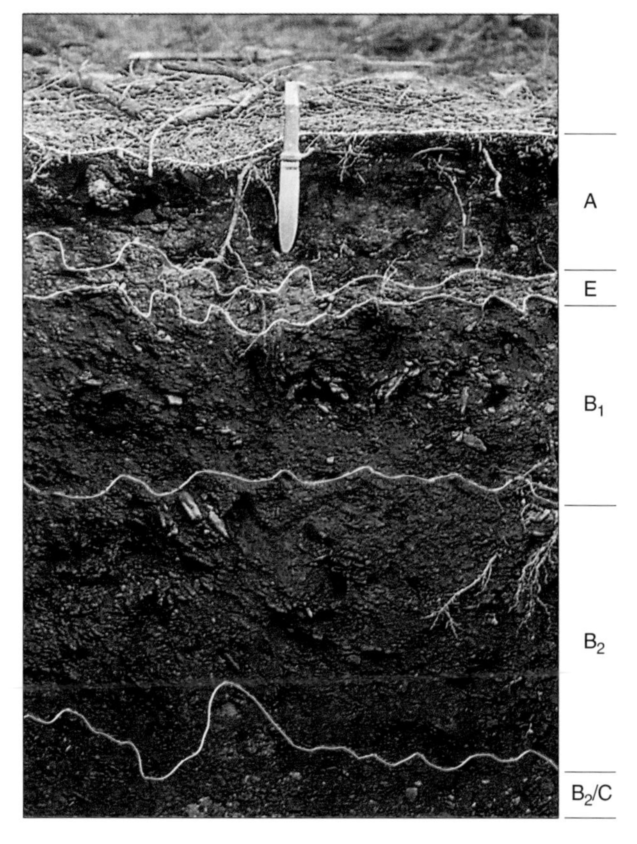

Figure 4.1 Profile of a spodosol with the different horizons identified.

with organic acids, such as fulvic acid, in the percolate. The aluminium and iron extracted from the A horizon is precipitated deeper down in the soil as aluminium and iron(III) hydroxides, $Al(OH)_3$ and $Fe(OH)_3$.

4.1.2 Weathering

Weathering involves the breakdown of primary minerals, such as feldspars and olivines, to form new secondary minerals. Aqueous ions are released into the soil water during this process. Weathering occurs when minerals in the soil come into contact with soil water. The process can be accelerated by the presence of CO_2 in the water (which lowers the pH of the water). Typical soil-weathering reactions are:

$$\underset{\text{potassium feldspar}}{4KAlSi_3O_8(s)} + 4CO_2(aq) + 22H_2O(l) = 4K^+(aq) + 4HCO_3^-(aq) + \underset{\text{kaolinite}}{Al_4Si_4O_{10}(OH)_8(s)} + \underset{\text{'soluble' silica (silicic acid)}}{8Si(OH)_4(aq)} \quad (4.1)$$

$$\underset{\text{olivine}}{(Fe,Mg)_2SiO_4(s)} + 4CO_2(aq) + 4H_2O(l) = 2(Mg^{2+}, Fe^{2+})(aq) + 4HCO_3^-(aq) + Si(OH)_4(aq) \quad (4.2)$$

$$\underset{\text{biotite}}{KMg_3AlSi_3O_{10}(OH)_2(s)} + 10H^+(aq) = K^+(aq) + 3Mg^{2+}(aq) + Al^{3+}(aq) + 3Si(OH)_4(aq) \quad (4.3)$$

$$CaCO_3(s) + CO_2(aq) + H_2O(l) = Ca^{2+}(aq) + 2HCO_3^-(aq) \quad (4.4)$$

The compositions of minerals vary from source to source. The parentheses in the formula for olivine above indicate that Fe and Mg cations may occupy a particular structural site in the mineral in varying proportions.

The secondary minerals formed by primary weathering reactions can further decompose in the soil solution; for example:

$$\underset{\text{kaolinite}}{Al_4Si_4O_{10}(OH)_8(s)} + 10H_2O(l) = \underset{\text{gibbsite}}{4Al(OH)_3(s)} + 4Si(OH)_4(aq) \quad (4.5)$$

$$Al(OH)_3(s) + 3H^+(aq) = Al^{3+}(aq) + 3H_2O(l) \quad (4.6)$$

The primary and secondary weathering processes release basic cations (such as Ca^{2+}, Mg^{2+}, Na^+ and K^+) and acidic cations (such as H^+ and Al^{3+}) into the soil solution.

4.1.3 Cation exchange

The cations produced by weathering reactions are not immediately transported away as the water drains from the soil. Instead, some cations are adsorbed onto the surface of the clay minerals and organic particles present in the soil, where they can be retained for long periods. The cation exchange capacity (CEC) of the soil is its ability to adsorb basic and acidic cations. This adsorption arises due to the net negative charge present on the surface of most soil particles that attracts and holds the positively charged cations in the soil solution.

○ Why do most clay minerals and organic particles in soil have a net negative charge?

● The charge comes from isomorphous substitution in clays (permanent charge), and dissociation of end hydroxyl groups, —OH, from clays and organic matter (pH-dependent charge).

The cations adsorbed to the surface of clays or organic particles are only bonded to the particles by weak electrostatic forces. They can thus be easily removed or exchanged by other cations in the surrounding solution (Figure 4.2).

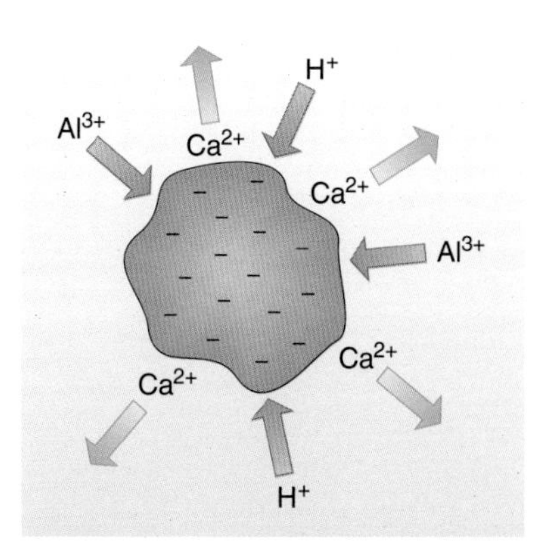

Figure 4.2 Schematic representation of cation exchange processes at the surface of a clay mineral or organic particle. In this case, the exchange is basic cations on the surface for acidic cations in the solution.

Cation exchange surfaces have differential preferences for basic and acidic cations. Exchanges between the basic cations Ca^{2+}, Mg^{2+}, Na^{+} and K^{+}, and the hydrogen ion, H^{+}, are of particular interest for the interactions of soils with acid deposition. These exchanges can be represented generically as:

$$\text{(clay–humus)}{-}M^{n+} + nH^{+}(aq) = \text{(clay–humus)}{-}nH^{+} + M^{n+}(aq) \qquad (4.7)$$

where M^{n+} represents any of the basic cations and n is the positive charge on that cation.

This is an equilibrium reaction and will shift from right to left (or vice versa) as the concentrations of M^{n+} and H^{+} ions in solution vary. The density of adsorbed basic cations on the particle surface thus depends on the characteristics of the surface and the relative amounts of acidic and basic cations in solution.

- ❍ What do we call the proportion of the total CEC charge that is occupied by basic cations at any one time, and in what units is it usually expressed?
- ● It is called the base saturation (BS) of the soil, and it is usually expressed as a percentage of the total CEC.

In many temperate regions, Ca^{2+} and Mg^{2+} are the most abundant basic cations on exchange sites as large numbers are produced by weathering (especially in areas with some carbonate rocks) and the relatively high affinity of these ions for the soil exchange surface. Although there is much variation, we can take the relative proportions of base cations making up the base saturation of a forest soil to be approximately:

Ca^{2+}	Mg^{2+}	K^{+} and Na^{+}
80%	15%	5%

In soils *not* affected by acid rain, the rates of the following processes are all in equilibrium (Figure 4.3a, overleaf):

- basic cation production by mineral weathering;
- basic cation exchange with the soil surface;
- basic cation loss by leaching from the soil.

The additional supply of hydrogen ions into a soil from acid deposition leads to the displacement of basic cations from the exchange sites on soil particles (Figure 4.3b). The cation exchange process therefore has the effect of lowering the concentration of hydrogen ions in the soil water (and in the soil drainage solution), and thereby neutralizing (in part, at least) the acid deposition that enters the soil. This neutralization has a cost, however. The soil particles themselves become acidified as a result of giving up basic cations and the base saturation of the soil is lowered.

If the input of hydrogen ions and their neutralization continues until almost all available exchange sites have been filled with hydrogen ions, the neutralization can no longer occur (Figure 4.3c). At this point, the base saturation of the soil approaches zero, and most of the H^{+} ions entering the soil from the atmosphere will pass through into the surface waters. As the pool of exchangeable base cations is exhausted, the flow of SO_4^{2-} passing through the soil will be balanced more and more by H^+ ions. This will increase the acidity of water leaving the soil, and lead to freshwater acidification (Figure 4.3c).

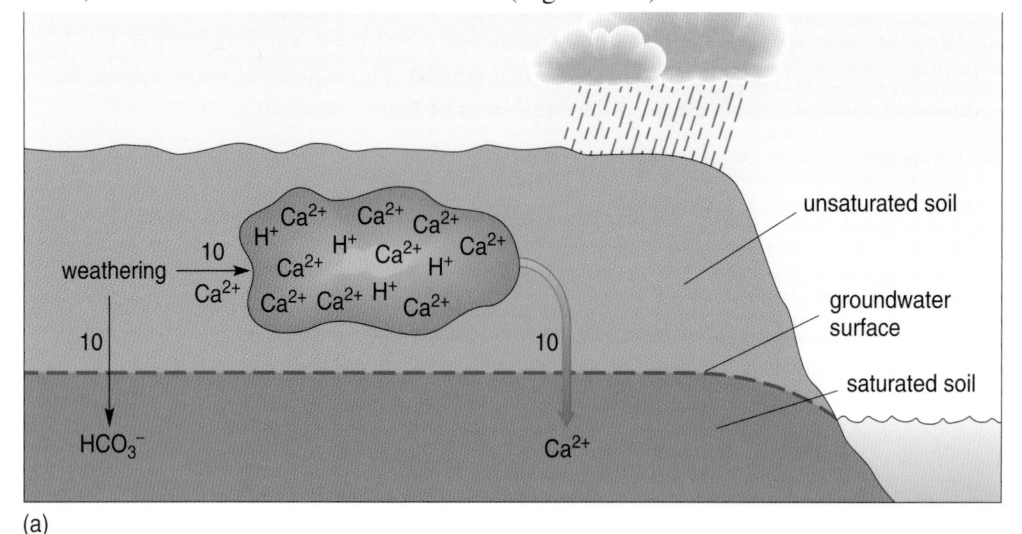

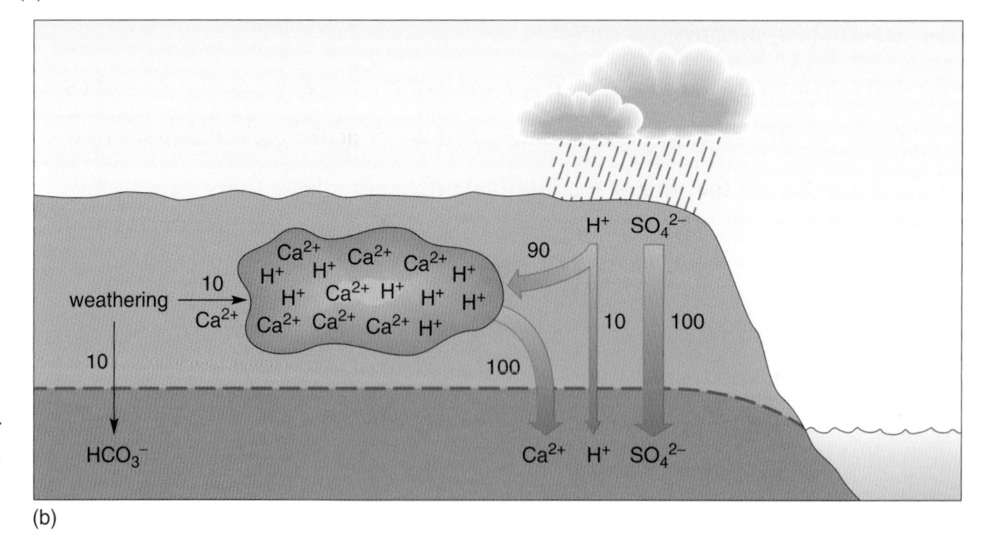

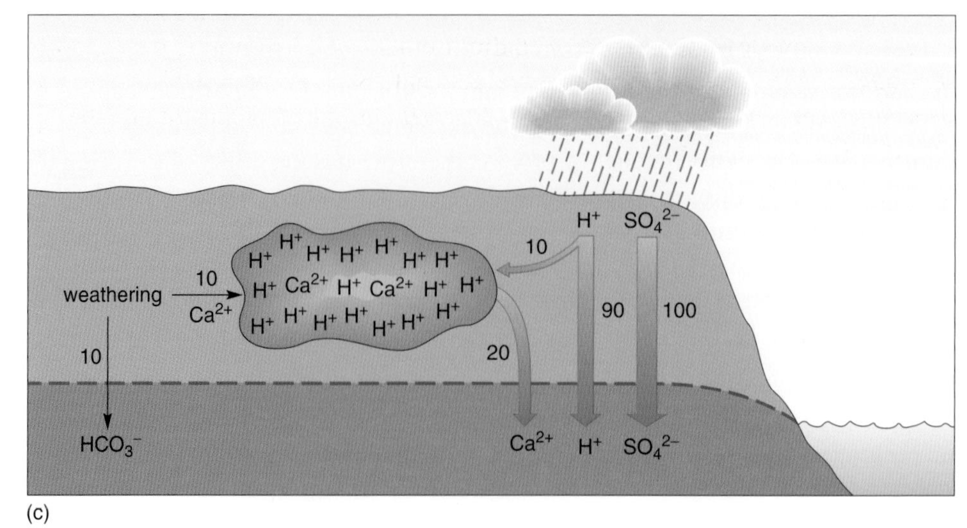

Figure 4.3 The production of basic cations by mineral weathering and the exchange of basic cations at the soil surface: (a) with no acid deposition; (b) in the presence of acid deposition percolating through a soil with high base saturation; and (c) in the presence of acid deposition percolating through a soil with low base saturation. The enclosed area in upper (unsaturated) soil represents exchange sites on soil particles. The numbers on the arrows represent hypothetical amounts of charge. Note that the total amount of positive charge exported in drainage waters (to the yellow zone at the bottom of each figure) balances the total amount of negative charge.

4.1.4 Sulfate adsorption

Another important process occurring in the soil that buffers soil water acidity from acid rain is sulfate adsorption. Some soils have the capacity to adsorb anions such as SO_4^{2-}. The capacity of a soil to adsorb and retain anions is known as its **anion exchange capacity** (AEC), and it is particularly important for soils with a high content of aluminium or iron oxides. The capacity of a soil to retain or release the anion sulfate, SO_4^{2-}, is known as its sulfate adsorption capacity (SAC).

Sulfate adsorption can occur through electrostatic attraction or complexation.

Through electrostatic attraction, sulfate ions are attracted to the protonated hydroxyl groups of clays with positive pH-dependent charges, such as aluminium or iron hydroxides:

$$\text{clay—(Fe,Al)—2}(OH_2)^+ + SO_4^{2-} \rightleftharpoons \text{clay—(Fe,Al)—2}(OH_2)^+(SO_4)^{2-} \quad (4.8)$$

The sulfate may subsequently form strong chemical bonds, or complexes, with the hydroxyl groups. These strongly bound anions cannot exchange as readily with ions in solution as can those anions that are held through simple electrostatic attraction.

A complexation reaction producing aluminium hydroxy sulfate at the soil–water interface can be represented as follows:

$$Al_2O_3(s) + 2SO_4^{2-}(aq) + 3H^+(aq) = 2Al(OH)SO_4(s) + OH^-(aq) \quad (4.9)$$

The process of sulfate adsorption counters soil water acidification in two ways:

- The formation of the iron or aluminium sulfate complexes consumes some hydrogen ions and releases some hydroxide ions. These complexes neutralize other hydrogen ions.
- The adsorption process limits the mobility of the sulfate ion in the soil solution, which, in turn limits the export of either acidic or basic cations in the drainage water.

Cations do not leach alone — they must be accompanied by an anion of equal charge. In general, all solutions must be in a state of 'charge balance'. This means that the total positive charges must equal the total negative charges for all ions in solution (Figure 4.3). Thus, if anions such as SO_4^{2-} are not exported in drainage waters, then acidic cations like H^+ cannot be exported either. The adsorption process can slow down the export of acidity from the soil by limiting the export of the **mobile anion**. Thus, the sulfate remaining in the soil through sulfate adsorption represents cations that are not leached out of the catchment into the surface waters downstream.

The maximum amount of sulfate that can be removed by the adsorption process will depend on the quantity and character of the aluminium and iron hydroxides in the soil.

❍ Where would you find soils with a high proportion of aluminium and iron hydroxides?

● Old, highly weathered, and/or highly leached soils, such as those in tropical areas, have a high proportion of aluminium and iron hydroxides.

For many soils, the SAC is relatively small, consequently the neutralization of H^+ and the retention of the mobile anion is also small. Even in those soils that initially have a high SAC, the year-to-year supply of SO_4^{2-} ions by acid deposition will eventually saturate the sulfate adsorption capacity of the soil. As the SAC is saturated, the rate of transport of SO_4^{2-} through the soil will increase.

The acidity of the water leaving a soil is therefore dependent on the amount of acid deposition on the soil, and the various acidifying and neutralizing processes that occur within the soil. These processes do not act independently, but influence one another. For instance, a soil with a high CEC and base saturation will have a large capacity for neutralizing incoming H^+ ions through exchange with base cations. The base cations released by this process will then be available to balance any SO_4^{2-} that has not been adsorbed. Clearly, soils with a large CEC and SAC have a large capacity to neutralize acid deposition. Conversely, acidification of the water leaving a soil will occur rapidly if the soil has a very low CEC, base saturation, and SAC. These soil properties are largely determined by the nature of the material forming the soils, and the processes of soil formation.

4.1.5 Mobilization of aluminium

Before we move on to Section 4.2, which covers the effects of these soil processes on the water in streams and lakes, we need to consider briefly the role played by aluminium in acidification. The most important ecological effect of the acidification of soils and soil water is that it releases dissolved aluminium into the soil solution, and ultimately into streams, rivers and lakes, and aluminium in some forms can be very toxic, particularly to fish. In the acid spodosols on resistant bedrock, such as those that occur in upland Britain and southern Scandinavia, calcium and magnesium are present in low concentrations, and the main exchangeable ion is aluminium. In the upper soil layers, aluminium tends to be complexed by the organic acids in the humus; in this form it is *not* poisonous to fish. However, when the pH becomes more acidic, this complexation of Al^{3+} is much reduced and inorganic aluminium ions are formed. It is aluminium in this inorganic form that is very toxic to aquatic life. At low pH, mineral particles such as gibbsite dissolve, producing aluminium cations:

$$Al(OH)_3(s) + 3H^+(aq) = Al^{3+}(aq) + 3H_2O(l) \qquad (4.10)$$

The Al^{3+} ions can react with hydroxide ions in the water to form a series of aluminium–hydroxyl species that are all in equilibrium with one another in solution:

$$Al^{3+} \rightleftharpoons Al(OH)^{2+} \rightleftharpoons Al(OH)_2^{+} \rightleftharpoons Al(OH)_3 \qquad (4.11)$$

The relative proportions of these different species depend primarily on the pH of the solution. All of these aluminium ions are toxic to fish in varying degrees, although Al^{3+} is generally considered the most toxic form.

Surveys show that the decline in fish stocks in Scandinavia correlates with the increasing acidity, increasing concentration of inorganic aluminium, and low calcium concentration of the lakes. Major fish kills have also been associated with short episodes of high acidity (brought on, for example, by winter storms or melting snow, when the pH of a stream can suddenly plummet from 6.0 to 4.6 in a matter of hours), when the concentration of inorganic aluminium is high and that of calcium is low.

4.1.6 Geological setting

Soils determine the capacity of a catchment to neutralize acid deposition, thus producing drainage waters that are not acidified. The soil properties that are critically important are those that determine its ability to consume incoming acids: primarily cation exchange and sulfate adsorption. These characteristics of soils are finite — all soils have a limited capacity for cation exchange or anion adsorption. Unless these soil properties are renewed, persistent acid deposition can result in a decline in the effectiveness of the neutralization processes, and an increase in the acidification of the catchment soils and waters. The rate of renewal of the key soil properties depends on the geological formations underlying the catchment soils.

The geological setting of the catchment thus determines, to a large extent, its long-term susceptibility to acidification by acid deposition. Those catchments underlain by rocks with relatively high weathering rates (such as carbonates and basalts) can produce large amounts of basic cations year after year. These cations, supplied to the soils of the catchment, can exchange for the acidic cations deposited from the atmosphere, thus neutralizing even large rates of acid inputs. Such areas are not susceptible to freshwater acidification.

Acidification is most likely to occur in areas of base-poor geology, such as those underlain by granite, quartzite, gneiss, or shales. These rocks weather slowly, so soil formation is slow and the soils tend to be immature. The soils also tend to be thin, with low base cation concentrations, CEC and SAC. In areas underlain by rocks with low inherent weathering rates, weathering cannot keep up with even moderate amounts of acid deposition. The inputs of acidic cations from the atmosphere (and subsequent exchange for basic cations in catchment soils) occur more rapidly than basic cations can be re-supplied by the primary and secondary mineral weathering processes. When this imbalance occurs, the catchment soils lose their ability to neutralize the incoming acids. Such areas are at risk of freshwater acidification.

4.2 Freshwater acidification

Section 4.1 demonstrated how soils could neutralize acid deposition, thereby preventing the acidification of surface waters. However, with increasing levels of acid deposition, the ability of a soil to neutralize H^+ and immobilize SO_4^{2-} may be exceeded, and H^+ and Al^{n+} ions will be released from the soil into surface waters such as streams, rivers and lakes. Sensitive surface waters will therefore begin to acidify.

A common measure of the ability of surface waters to resist acidification is the **acid neutralizing capacity** (ANC). The ANC is simply the sum of the molar concentrations of all the ions (proton acceptors) that can potentially neutralize H^+ ions, minus the molar concentrations of all the ions (OH^- acceptors) that can potentially neutralize OH^- ions and the molar concentrations of all the ions that can potentially release protons (proton donors) in the water, weighted by charge. In most natural waters:

$$\mathrm{ANC} = [\mathrm{HCO_3^-}] + 2[\mathrm{CO_3^{2-}}] + [\mathrm{OH^-}] + [\mathrm{org^-}] - 3[\mathrm{Al^{3+}}] - 2[\mathrm{Al(OH)^{2+}}] - [\mathrm{Al(OH)_2^+}] - [\mathrm{H^+}] \quad (4.12)$$

where the square brackets represent the concentration of each ion in, for example, μmol per litre, and 'org^-' is the net negative charge on all dissolved organic

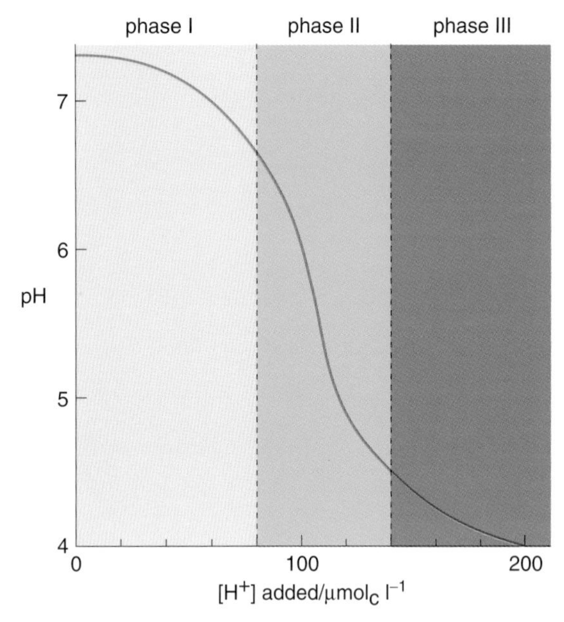

Figure 4.4 This curve shows the addition of 200 μmol_c (μmol of charge) of hydrogen ion, H^+, to a litre of carbonate solution with an initial HCO_3^- concentration of 100 $\mu mol_c\, l^{-1}$. The acid neutralizing capacity of this solution (ANC = 100 $\mu mol_c\, l^{-1}$) prevents the pH from declining below 6 until half of the acid has been added. The ANC of natural waters similarly prevents the occurrence of low pH values as long as the amount of added acid is less than the ANC. Phases I, II, and III roughly approximate stages of acidification in sensitive surface waters.

compounds in the water. Incoming H^+ ions are neutralized by the proton acceptors, that is the negative ions, until all the proton acceptors in the solution are used up, at which point the acidity of the water will increase. In natural waters close to neutral pH (in the range pH 6–8), the principal proton acceptor is hydrogen carbonate, HCO_3^-. The hydrogen carbonate in the water reacts with hydrogen ions to form carbon dioxide and water:

$$HCO_3^-(aq) + H^+(aq) = H_2O(l) + CO_2(g) \qquad (4.13)$$

❍ Where does the hydrogen carbonate come from?

● It comes from the weathering of primary minerals in the presence of dissolved carbon dioxide (Equations 4.1–4.4).

In most natural waters, with low levels of dissolved organic matter (org^-), the pH will remain stable until the supply of hydrogen carbonate is exhausted, at which point further input of H^+ ions will lead to rapid acidification (Figure 4.4).

In catchments where calcium and magnesium are important components of soils, for example those in which rocks are dominated by easily weathered minerals such as carbonates and feldspars, there is usually enough ANC generation to neutralize acid deposition. When catchment soils have low base cation concentrations and low weathering rates, the release of hydrogen carbonate into freshwaters is low and ANC generation is limited. Acidification of surface waters in such catchments can take place relatively easily through acid deposition and the release of H^+ ions from soils.

❍ In addition to the amount of calcium and magnesium in soils, what other important factor determines whether the water percolating through a catchment is ion-rich or ion-poor?

● Catchment hydrology: the flowpath, the amount of surface flow compared to throughflow or groundwater, and the rate of flow of water. These factors affect the reaction time between the water and the soil particles.

Acidification in the ion-poor ('soft-water') lakes that are typical of the uplands of Britain and southern Scandinavia can therefore be seen as a three-phase process, very roughly approximated by the laboratory-derived curve shown in Figure 4.4.

- In phase I, the dissolved hydrogen carbonate counters the effects of increasing H^+ input from the atmosphere. In this stage, pH will remain above 6, and there will be little effect on plant and animal communities.
- Phase II occurs when the hydrogen carbonate is almost consumed. During this period the pH of the water is unstable and can decline dramatically during periods of heavy inputs of acid deposition, such as in winter storms. These increases in acidity can lead to significant biological effects, such as fish kills.
- The third and final phase of acidification occurs when the ANC is depleted. The pH of the water falls rapidly to low levels, typically below pH 5.0, but begins to stabilize.

4.3 Summary of Section 4

1. The main characteristics of soils that determine their ability to counter acidity are their cation exchange capacity (CEC), base saturation (BS), and sulfate adsorption capacity (SAC).
2. The acid neutralizing capacity (ANC) is a measure of the ability of surface waters to resist acidification. It is calculated as the sum of the concentrations of all of the proton acceptors minus the concentrations of all the OH^- acceptors and proton donors.
3. The CEC and BS of soils, and the ANC of surface waters, are strongly related to the catchment geology. Catchments dominated by carbonates show a high CEC, BS and ANC due to the release of basic cations and HCO_3^- into the soil solution from weathering.
4. The SAC is related to the soil type, with soils dominated by aluminium and iron hydroxides showing a high SAC.
5. The process of sulfate adsorption through complexation counters soil acidification by consuming hydrogen ions. Sulfate adsorption also limits the mobility of sulfate that, in turn, reduces the export of cations into drainage water.
6. Base cations, both in the soil water and adsorbed onto the surface of soil particles, can exchange with H^+ ions in the soil water. This lowers the acidity of soil water but increases the acidity of the soil particles.
7. H^+ ions in less acidified soil solution are neutralized by HCO_3^- ions to form CO_2 and H_2O. This is particularly effective in carbonate-dominated catchments.
8. In more acidified soils, Al^{3+} ions on soil particles exchange with the H^+ ions in the soil solution. At low pH, Al–hydroxyl species released in solution can be very toxic to fish.

Question 4.1

A particularly dangerous time for fish in streams is in the spring, when snowmelt or runoff can bring a major flush of acidity into streams, just at the most sensitive stage of many fish species. Why is this water especially acidic?

Question 4.2

It is often observed that lakes and streams receiving highly organic runoff, such as those draining peatlands, support populations of invertebrates, fish and other biota that are absent from clear lakes receiving the same amount of acid deposition but no organic runoff. Give at least one explanation for this observation.

Question 4.3

In your answer to Question 3.1, you determined that mountainous areas receive more acid deposition than lowland areas. In the light of some of the topics in Section 4, why might mountainous ecosystems, in addition to *receiving* more acid deposition, be more *sensitive* to acidification through acid deposition?

5

Lake acidification

In the previous sections we have seen how anthropogenic sulfur emissions cause acid deposition, which, in turn, can bring about the acidification of soils and surface waters. Even though the potential acidifying effects of sulfur compounds on freshwaters have been recognized for some time, there has still been intense scientific and political debate as to whether acid precipitation in general, and sulfur deposition in particular, is responsible for the widespread observations of low-pH lakes and streams, and the declines in fish stocks discussed in Section 1. This debate took place because, until the mid-1980s, the evidence linking acid deposition to lake acidification lacked a temporal perspective.

The lack of good, long-term records of measured lake water acidity in the affected regions meant that there was no absolute proof that lakes and streams in these areas had acidified over the last few decades. It was also difficult to demonstrate categorically that, if acidification had taken place, the culprit was acid deposition. Consequently, alternative hypotheses were suggested to explain the observations of acid lakes and streams. This section will review these hypotheses, and then discuss the research that demonstrated the clear cause-and-effect relationship between acid deposition and lake acidification. Although acidification can be seen as a problem of environmental chemistry, the research that confirmed the importance of acid deposition was primarily based on biological and geological techniques.

5.1 Possible causes of lake acidification

In the early 1980s some scientists argued that freshwater acidification had *not* taken place in southern Norway or Britain, and those lakes with low pH may always have been acidic. However, the decline in fish stocks observed in rivers and lakes in both Scandinavia (see Section 1) and in areas of Britain such as Galloway, Scotland suggested that some chemical changes had taken place in these surface waters. Most scientists therefore accepted that acidification had taken place, but three competing sets of hypotheses were proposed to explain the acidification of surface waters:

- natural long-term acidification;
- changes in land use;
- acid deposition.

5.1.1 Natural long-term acidification

This hypothesis proposed that surface waters in upland areas acidify naturally over long periods of time due to changes in catchment soils and vegetation. For instance, plant roots and micro-organisms release carbon dioxide by respiration, which dissolves in the soil water and dissociates to produce hydrogen ions and hydrogen carbonate ions (Equation 3.2). The roots of plants and trees take up basic cations from the soil and release hydrogen ions to maintain charge balance. The basic cations are partly returned to the soil by leaching from the leaves and by natural decomposition, but the net effect is a gain of hydrogen ions in soil solution. Organic acids such as humic and fulvic acids are released by the decomposition of plants. These organic acids dissociate, releasing hydrogen ions.

This hypothesis predicts that a slow rate of acidification has taken place over the past 10 000 years since the retreat of the last ice-sheets. Indeed, this process of long-term soil acidification is known to occur in Scandinavia, Britain and North America. It has been suggested that the contemporary acidity of some very acid upland lakes in the Lake District was the result of this process, and that since 1800 no enhanced acidification has taken place.

5.1.2 Changes in land use

The land-use hypothesis argues that recent acidification of lakes and streams has taken place, but that the acidification was due to changes in catchment management and land use over the last century. Heathland regeneration and conifer afforestation in upland areas were both proposed as causes of the acidification.

The 'heathland regeneration' hypothesis argues that acidification is due to the decline of upland agriculture. A decrease in the intensity of farming in the uplands would lead to a corresponding decrease in grazing pressure and a decline in the regular burning of upland grazing land. Such changes would result in the decline of grass communities and the regeneration of heathland vegetation such as heather. Heathland vegetation would increase the amount of acid organic material in the soil, and release H^+ ions into surface waters. These processes could all lead to lake acidification.

The 'afforestation' hypothesis is based on the fact that over the last 80 years there has been an extensive programme of conifer afforestation in the uplands of the UK. In Scotland, comparisons between streams in afforested and non-afforested catchments indicate that the former have higher sulfate values and lower pH values. The suggestion is that conifer afforestation per se was responsible for recent surface water acidification, through one or a combination of the following three mechanisms:

- acid production by oxidation of anaerobic soils or peats, following ploughing for planting trees (with ploughing exposing deeper anaerobic soils to the air);
- depletion of soil base cations through uptake by trees, following afforestation;
- accumulation of acid humus in the soil from the forests.

5.1.3 Acid deposition

The final hypothesis is that acidification of surface waters has occurred in the last 150 years as a result of acid deposition resulting mainly from anthropogenic emissions of sulfur dioxide. This hypothesis predicts that acidification took place following the increase in fossil fuel burning and industrial emissions during the Industrial Revolution.

5.2 Resolution: palaeolimnology and lake acidification

Each of the hypotheses proposed to explain the observations of acidic rivers and lakes predicts that acidification occurred under different conditions, at different times and at different rates. The problem is: how can we design a test to determine which hypothesis is most likely to be correct? Such a test requires historical data on water quality change that can be compared to changes in land-use practice and can also be assessed for the timing and rate of any change. Yet no conventional long-term water quality records are available for sites believed to have acidified.

Figure 5.1 The Round Loch of Glenhead in Scotland.

The solution to this problem was provided by the mud lying at the bottom of lakes. One lake in the UK, the Round Loch of Glenhead in Scotland (Figure 5.1), played a pivotal part in this solution.

Lake sediment consists of various components, including:

- organic material produced within the lake by plant growth;
- inorganic sediments such as sand and silt washed into the lake from the catchment;
- material derived from the atmosphere.

The sediment accumulates continuously through time, providing an archive of material relating to the history of the lake and its catchment. The study of lake sediments and the history of lakes is called **palaeolimnology**.

The skeletal remains of diatoms are the components of lake sediments that palaeolimnologists are most interested in. Diatoms are microscopic, single-celled algae that are abundant in nearly all aquatic systems (Figure 5.2).

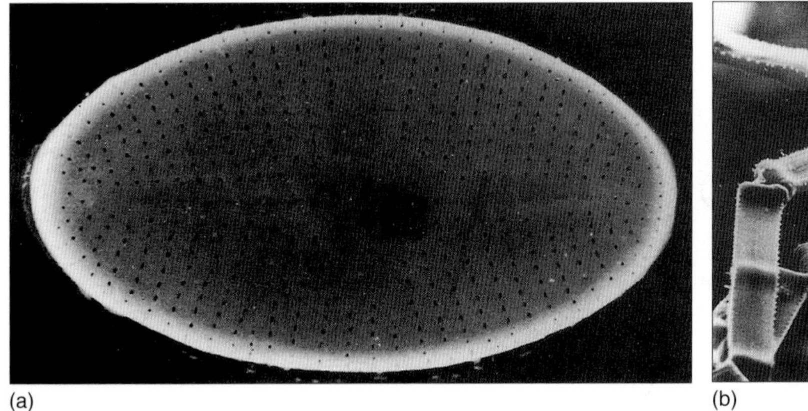

(a)

(b)

Figure 5.2 Scanning electron micrographs of two species of diatom commonly found in freshwater lakes, and their preferred pH ranges: (a) *Achnanthes minutissima* (pH 5.0–5.5; magnification ×113 000); (b) *Tabellaria quadriseptata* (pH < 5.0; magnification ×1750).

Diatoms have three characteristics that make them invaluable for studies of lake acidification:

1. Their cell walls are made of resistant silica. After death, the diatom remains accumulate in the lake sediments, where they are preserved. Lake sediments, therefore, contain a record of the diatoms that lived in the lake at different stages of the lake's history.
2. Diatoms are highly responsive to changes in water quality.
3. Different species of diatoms prefer different conditions of water chemistry.

The combination of different diatom species found in a lake will therefore be influenced by the water chemistry. In soft-water lakes, diatoms are excellent indicators of lake water pH. Mathematical models have been developed that link diatom assemblages collected from lakes to measured pH values. These models indicate that the change in the pH of the lake water can be predicted with a high degree of accuracy from the change in the relative proportions of the remains of different diatom species that were deposited at the time.

Palaeolimnologists can reconstruct the pH history of a lake from the diatom remains preserved in the lake sediments. Cores of sediment are retrieved from a lake and sliced into sections. Chemical techniques are used to calculate the date at which each slice of sediment was deposited. The diatoms are extracted from each slice, and the relative abundance of different species enumerated by microscopic examination. Variations in the structure of the diatom cell walls are used to distinguish between individual species. The diatom assemblage present in each slice is then used to infer the pH of the lake *at the time of deposition*. The acidification history of the lake can be reconstructed by studying the diatoms in successive layers of mud from the bottom of the lake.

Such reconstructions of past lake water pH allow the timing and rate of acidification to be assessed. Diatom-based palaeolimnological studies can be used in this way to test the hypotheses of acidification described in Sections 5.1.1 to 5.1.3.

The Round Loch of Glenhead has a contemporary pH of 4.8. It is located in Galloway, southwest Scotland, an area where many lakes have current pH values of less than 5.0. A sediment core, representing a time-span of approximately 150 years was retrieved from Round Loch in 1989. Analysis of the historical lake pH values represented by the diatom assemblages in the core demonstrated that, over the last 130 years, the loch had acidified by nearly one pH unit (Figure 5.3).

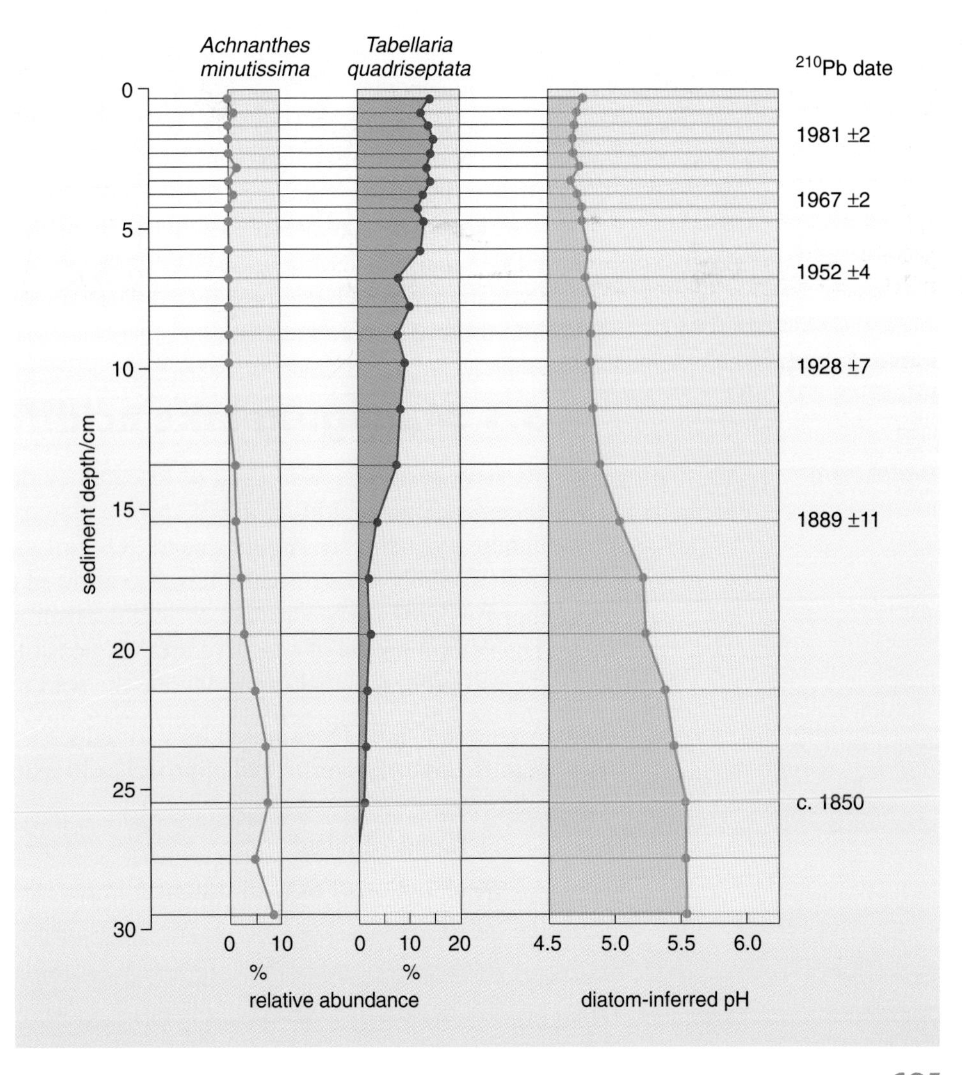

Figure 5.3 Relative abundance curves (% of 15 species in total) for the acid-sensitive and the acid-tolerant diatom species pictured in Figure 5.2, and reconstructed pH curve for a sediment core from the Round Loch of Glenhead, Galloway, Scotland. The pH reconstruction was based on 15 species. The left-hand axis represents the depth in the core. The dates corresponding to each depth were determined by ^{210}Pb dating and are given on the right-hand axis.

❍ What change in acidity does the acidification of the loch by one pH unit represent?

● A tenfold increase in acidity.

So what has caused this acidification in the Round Loch of Glenhead? The acidification is recent and relatively rapid. It began slowly after 1850 and then accelerated through the late 19th century and early 20th century. Long-term acidification due to slow post-glacial build-up of organic matter is therefore *not* responsible. Additionally, land-use changes cannot have caused the acidification. This is because the catchment of the loch has not been afforested, and records show that there has been no reduction in the intensity of burning or grazing over the past 200 years. The only plausible explanation for the acidification is acid deposition. The timing of the first pH change is consistent with the acid deposition hypothesis, since it occurred around 1850 when air pollution levels in the UK are thought to have started to rise due to industrialization. The pH dropped sharply around the turn of the century, when industrial pollution accelerated.

Studies such as this one from the Round Loch of Glenhead have been repeated at many sites across Britain and Scandinavia. These studies consistently demonstrate that lakes in areas of thin soil, overlying granite or gneiss, where levels of acid deposition are high, have acidified in industrialized countries over the last 150 years. The acidification is more recent in Scandinavia as the tall smokestacks that allow long-distance transport of these pollutants from countries like the UK and Germany have only been around for about the last 40 years. There is no evidence that heathland regeneration has caused acidification in Britain. Palaeolimnological studies have also shown that afforestation by itself does not cause lake acidification, but that forests in areas of high acid deposition may exacerbate acidification due to the increased scavenging of sulfur pollutants from the air, which can be dry-deposited on leaves or needles and later washed into the soil in precipitation.

5.3 Managing lake acidification

We have seen that, as a result of research evidence, there is widespread acceptance that acidification of surface waters has taken place over the last hundred years or so, largely due to SO_2 emissions produced during the burning of fossil fuels. That research has also allowed countries to identify those areas that are particularly sensitive to acid deposition (e.g. for the UK in Figure 5.4). The determination of sensitive areas is possible because of our understanding of the soil processes that can moderate the acid inputs.

Attention is now focusing on how to reduce and eliminate the problems of lake acidification. Two principal approaches to managing the problem have been advocated:

- the liming of lakes
- the reduction of SO_2 emissions.

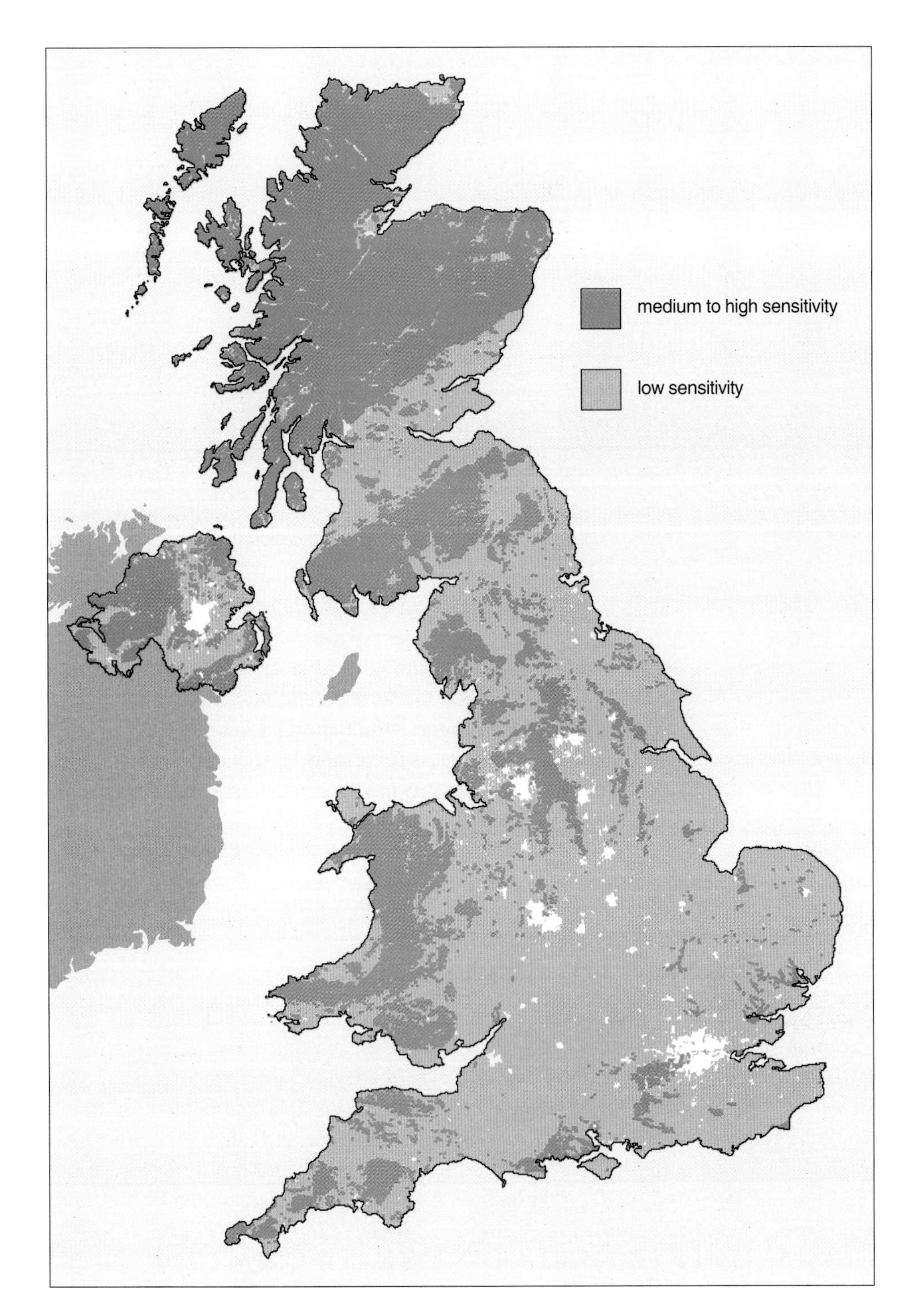

Figure 5.4 Map of the UK showing areas sensitive to acid deposition, based on the characteristics of soils, rocks and surface waters. (White areas are urban areas or areas where information is missing.)

5.3.1 Liming of lakes

In Norway and Sweden, it has been shown that treating an acid lake with powdered limestone can quickly increase the pH of the water (Figure 5.5). A project in the UK, based at Loch Fleet, has investigated liming the surrounding land rather than the lake itself. Theoretically this should result in a slower release of lime to the lake, and hence longer lasting effects. Over 3000 tonnes of limestone were applied to the land in April 1986, with more added in 1987. The improvements persisted to such an extent that it was possible to reintroduce brown trout to the lake in 1987.

Figure 5.5 A large number of lakes in Sweden have been limed to counteract the effects of acid deposition. Lake liming is also employed in other countries (including the UK), but to a much lesser extent than in Sweden.

These procedures are not without their own problems. First, liming is very expensive, and it often needs repeating regularly to maintain the pH improvement. It is also difficult to judge the amount of lime required to restore the lake to its pre-acidification acidity — after liming, Loch Fleet became more alkaline than it had ever been before in its history. This alkalinity can adversely affect the surrounding flora and fauna, which have adapted over the years to the local conditions. Liming, therefore, is a very drastic remedy that may be justifiable in unusual circumstances, such as the preservation of a threatened species, but that is never going to become a universal solution to acid deposition. For that, it is necessary to tackle the root cause.

5.3.2 Reduction of SO_2 emissions

The clear cause–effect relationship between acid deposition and lake acidification suggests that the most obvious method of controlling acidification is to reduce acid deposition (Figure 5.6). Attention in acid rain research is therefore focusing on the reductions in sulfur emissions and acid deposition that would be required to reverse acidification and allow affected lakes to recover. However, because air pollution generated in one country can lead to deposition in another country, international agreements are needed to reduce levels of acid deposition in certain regions. For example, a large proportion of acid deposition in Norway originates from other countries, especially the UK and Germany. Reductions in SO_2 emissions are required in these other European countries to reduce acid deposition in Norway.

Negotiations to reduce sulfur emissions have been conducted on a European scale under the auspices of the United Nations Economic Commission for Europe (UNECE). The Scandinavian countries had persuaded UNECE to promote protocols for reducing SO_2 emissions across Europe, even before there was clear scientific evidence to demonstrate categorically the link between sulfur emissions and acidification. In 1985 many European countries adopted the first 'sulfur protocol', which encouraged member states to reduce national emissions of SO_2 by a minimum of 30% by 1993, based on reference emissions in 1980. Several net sulfur-exporting countries failed to sign the protocol, including Poland and the United Kingdom.

Subsequently, UNECE member nations have met several times under the convention for Long-Range Transboundary Air Pollution (LRTAP) and agreed to several protocols for the control of emissions. Under LRTAP, the UK is party to:

- the 1988 Sofia Protocol concerning the control of emissions of nitrogen oxides or their transboundary fluxes, and
- the 1994 Oslo Protocol on further reductions of sulfur.

The UK is also a signatory to the 1999 Gothenburg Protocol on abatement of acidification, eutrophication and ground-level ozone.

For the Sofia Protocol, the UK agreed to control and/or reduce emissions of nitrogen oxides, or their transboundary fluxes, so that, by 31 December 1994, they did not exceed their 1987 emission levels. In addition, the UK agreed to apply emission standards to major new stationary sources, such as new power plants, and to mobile sources of major source categories such as new cars, and to introduce pollution control measures for existing major stationary sources.

For the Oslo Protocol, the UK agreed to the schedule of SO_2 emission ceilings in kilotonnes per year (kt yr^{-1}) shown in Table 5.1.

Figure 5.6 Reduction of emissions produces appreciable improvements in air quality and, ultimately, ecosystem health. This photograph shows natural regeneration of young spruce trees in an acid rain-damaged forest in the Karkonoski National Park, southwest Poland. Up until the 1990s, these trees were subject to severe acid precipitation, which was wind-blown from the Czech Republic.

Table 5.1 SO_2 emission ceilings agreed in the UK for the Oslo Protocol.

	Year		
	2000	2005	2010
SO_2 emission ceiling/kt yr^{-1}	2499	1470	980
% decrease in SO_2 emissions compared to 1980 emissions	50	70	80

In addition, the UK agreed to apply emission limits to major new stationary sources and existing sources by 1 July 2004 (not entailing excessive costs), and to apply national standards to the sulfur content of gas and oil.

For the Gothenburg Protocol the UK's obligations are to achieve the emission ceilings shown in Table 5.2 by the year 2010; note the more stringent limit for SO_2 emissions.

Table 5.2 Emission ceilings for 2010 agreed in the UK for the Gothenburg Protocol.

	Sulfur dioxide	Nitrogen oxides	Ammonia
Emission ceiling/kt yr^{-1}	625	6671	3129
% decrease in emissions compared to 1990 emissions	83	49	15

You can see from Tables 5.1 and 5.2 that Gothenburg tightened the existing controls.

The adoption of these protocols has brought about a major reduction of SO_2 emissions in Europe (for example Figure 2.2b), and in some areas terrestrial and aquatic ecosystems are showing signs of recovery (Figure 5.6). However, nitrogen pollutants, especially ammonia, are being reduced to a much lesser degree — and are even increasing in some areas. These compounds are deposited on ecosystems as NO_3^- and NH_4^+, and can contribute to ecosystem acidification, potentially delaying recovery from sulfur-based acidification. Emissions of nitrogen compounds (for example from car exhaust gases, and agriculture) are more difficult to control than sulfur sources as they are many and diffuse rather than individual, identifiable sources as is the case with sulfur (such as large power plants).

SO_2 pollution, and the resulting acid rain, is an increasing problem in Asia (Figure 2.2a), where industrialization is increasing at a rapid pace without the strict emission controls that have been adopted in North America and Europe.

5.4 Summary of Section 5

1. The pH of a lake at the time a sediment was deposited can be determined from the fossil diatom assemblages present.
2. Evidence shows that some lakes have acidified by tenfold since the 1850s.
3. Afforestation exacerbates acidification but is not the primary cause.
4. Palaeolimnology has shown that sulfur dioxide emissions are the main cause of acid rain.
5. Some acidified lakes have been treated with lime, applied either directly into the water or on the surrounding catchment area, to reduce the acidity. pH levels recover very quickly, but the procedure is expensive, may adversely affect sensitive biota, and needs repeating regularly to maintain the pH improvement.
6. SO_2 emissions have been reduced in developed countries through the adoption of international protocols, but it is uncertain how long it will take lakes affected by acid deposition to recover. SO_2 emissions are increasing in Asia and many developing countries.
7. Some of the recovery from acid deposition in Europe may be delayed by continued high and, in some cases, increasing levels of nitrogen pollution, especially NO_3^- and NH_4^+.

Question 5.1

If surface water acidification were a natural process, with surface water pH declining slowly from, say, a near-neutral pH over the last 10 000 years since the retreat of the last ice sheets, how might the relative abundance of the two diatom species in the sediment core depicted in Figure 5.3 look? How would the pH reconstruction from the same core be different from that shown in Figure 5.3?

Modelling the effects of acid deposition

6

As we have seen above, research has focused attention on certain physical and chemical processes in the soils of catchments as keys to the responses of surface waters to acidic deposition. These processes include:

- weathering of minerals in catchment soils as a source of basic cations (calcium, magnesium, sodium, potassium);
- exchange of basic and acidic cations by catchment soils;
- anion retention by catchment soils (e.g. sulfate adsorption);
- dissolution of aluminium hydroxide compounds by reaction with hydrogen ions.

Critical decisions about the management of our ecosystems depend on how quickly, and to what extent, these processes affect the responses of surface waters to changes in acidic atmospheric deposition. Surface water acidification and recovery have occurred (and will occur) over time-scales of years to decades in natural systems. If we are to plan and manage future emissions of SO_2 in order to reduce or alleviate surface water acidification, we must use existing information to predict the patterns, time-scales and magnitudes of long-term changes in surface water quality in response to actual or assumed changes in the acid deposition. Mathematical models based on our understanding of the chemical and physical processes operating in soils and surface waters are used for this purpose.

In this section, we shall briefly examine the ways in which these models allow us to understand the long-term patterns of soil and water acidification that occur as acid deposition increases, and to predict the long-term patterns of recovery that might be expected as acid deposition declines.

6.1 Acidification models

Models of acidification are usually based on principles of conservation of mass within ecosystems. That is, the difference between what comes into soils (from atmospheric deposition) and what flows out of soils (the discharge into lakes and streams) must be accounted for within the soils themselves. Thus, the difference between the acidity of atmospheric inputs and the acidity of surface water drainage must be quantifiable in terms of chemical processes within the soil. The most important effects of acidic deposition on catchment surface water quality are:

- decreased pH and acid neutralizing capacity (ANC), and
- increased basic cation and aluminium concentrations.

The soil reactions discussed in earlier sections can be used to understand the patterns of these changes, and to build mathematical models of the acidification process.

A number of such models of soil and surface water acidification were developed in the early 1980s. These models were all based on the same conceptual understanding of soil processes. They differed primarily in the extent to which they attempted to represent the spatial variations in soil processes that occur within catchments, and the time-scales of the inputs and outputs of each model.

These differences in detail reflect the different purposes for which the models were built, ranging from estimating transient water quality responses for individual storm events to estimating chronic acidification of soils and base-flow surface water.

As new information about the effects of acid deposition became available over the years, some of the models were abandoned as inconsistent with the new observations. Others were modified in the light of the new knowledge and continued to be used. One of these models, named MAGIC (Model of Acidification of Groundwater In Catchments), has been in use since 1985. MAGIC has been applied extensively in North America and Europe to both individual sites and regional networks of sites, and has also been used in Asia, Africa and South America. It is the primary model upon which policy and assessment activities on the effects and mitigation of acid rain are based for European and North American countries.

To illustrate how acidification models work, and the information they impart, we will examine two different examples of how the MAGIC model has been used in studies of acid deposition effects. The first is an experimental application designed to answer some general questions about the time-scales of acidification. This study is scientific in nature and is designed to generate understanding of the acidification process. The second example is an applied study designed to evaluate the effects of the latest agreed reductions of SO_2 emissions on lakes in sensitive areas of the UK. This kind of model application is more management-oriented and attempts to use our scientific understanding to assist in the formulation of policy regarding protection of environmental resources.

6.2 Experimental model

A hypothetical catchment is used in this experiment. The catchment is sensitive to acid deposition, but not excessively so. It has low ANC prior to the onset of acid deposition, so that it has the potential to acidify. The soils have a moderate sulfate adsorption capacity and a moderate base saturation so that a relatively slow response should result when the acid deposition starts. The slow response allows us to examine the different stages of acidification and recovery that are possible, given the complexity of the inter-linked soil processes. While not representing any particular catchment in the real world, the soils' characteristics of the hypothetical catchment are similar in many aspects to those of moderately sensitive catchments in the UK (Figure 5.4).

The acid deposition inputs used in this experiment are highly simplified. The first years of the experiment have no acid deposition, and the model output represents that of an undisturbed system. At year 1, the acid deposition is suddenly switched on, and is held constant for 120 years at a level typical of deposition in central UK in the early 1990s. At year 121, the acid deposition is switched off, and the model is run for an additional 130 years to observe the recovery of the hypothetical catchment.

This simple 'on–off' input of acid deposition is, of course, at odds with our knowledge of how acid deposition actually developed over many decades (Figure 2.2), and with current plans to reduce emissions and thus acid deposition in the future (Section 5). However, the ability to simulate simple changes in acid deposition in order to observe the responses of a complex system is one of the most useful aspects of mathematical models. Using a simple 'on–off' input of acid deposition allows us to estimate the overall response times of what is a highly complex set of interacting and competing soil processes.

6.2.1 Output of the model experiment: acidification and recovery

The results of the modelling experiment for a stream draining the catchment (Figure 6.1) are divided into the stages of pre-acidification (Stage 1), acidification (Stages 2–4) and recovery (Stages 5–7) as a convenient means of comparing the time-scales of changes in water quality.

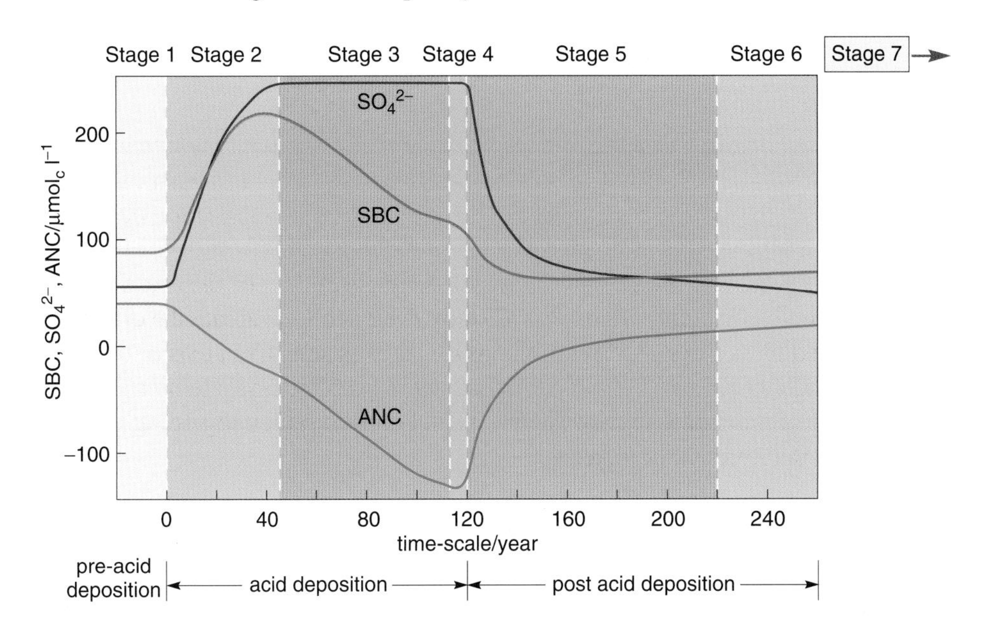

Figure 6.1 Predictions of the MAGIC model for the experimental simulation. The term SBC refers to the sum of basic cation concentrations, and ANC is acid neutralizing capacity. The model simulates the individual responses of the basic cations Ca^{2+}, Mg^{2+}, Na^+ and K^+ to the changes in acid deposition. For clarity, only the total of the four cations is shown here. Concentrations of SBC, sulfate ion and ANC in the stream water are all in μmol$_c$ l^{-1}.

Stage 1 Pre-acidification

This is the steady state prior to the increase in atmospheric deposition of sulfur. Recovery of the hypothetical catchment after deposition ceases is assessed by the rate at which the water quality variables return to these pre-acidification levels.

Stage 2 Sulfate adsorption

During this stage there is a lag in the increase of sulfate ion concentrations in the stream (Figure 6.1) as the soils adsorb the atmospherically deposited sulfur. This sulfate slowly fills the adsorption sites, with the result that sulfate ions in the drainage water slowly increase. As sulfate concentrations in the stream slowly increase, basic cation concentrations also increase (SBC in Figure 6.1 is the sum of all basic cations, that is Ca^{2+}, Mg^{2+}, Na^+ and K^+). This increase in base cations is the result of cation exchange reactions in the soil (Figure 4.3). However, the increase in basic cations is not keeping pace with the increase in sulfate, causing a decline in the acid neutralizing capacity (ANC). Given the sulfate adsorption characteristics of the hypothetical catchment, 40–50 years must elapse before the soil adsorption sites are filled and the stream sulfate concentrations stop increasing. In this hypothetical catchment the ANC is depleted (ANC 0) after approximately 20 years. The rate of supply of base cations from soil cation exchange reactions reaches a peak after approximately 30 years, as indicated by the plateau and downward turn in base cation concentrations. The response time for the initial acidification of the catchment is in this case about 45 years.

Stage 3 Saturated sulfate adsorption capacity

Sulfate concentrations reach a new steady value as the soil sulfate adsorption sites are filled. The flux of high concentrations of sulfate through the soil continues to deplete the store of exchangeable base cations. Even though the catchment reaches a steady value of sulfate in 45 years (Stage 2), base cation concentrations continue to decline for many years afterwards.

Stage 4 Base cation depletion

Given continued acid deposition, base cation concentrations should eventually return to pre-acidification levels once the base saturation of the soils has been depleted (Figure 4.3) because at this stage basic cation levels are controlled only by the primary mineral re-supply rate. Because we have truncated Stage 4 in this example by stopping acid deposition, that level of acidification is not reached in this example. However, the rates of change of base cation concentrations (and ANC) decrease markedly near the end of Stage 3 due to depletion of basic cations from soil exchange sites.

Stage 5 Sulfate desorption

This stage begins when acid deposition decreases to the initial low level. The rate of decline of stream water anion concentrations will depend on conditions in the soil because the sulfate adsorbed by the soil must be flushed from the system. As the anion concentrations decrease, the base cation concentrations will also decrease and ANC will begin to increase as fewer hydrogen ions are needed to balance the strong acid anions. The acidification and recovery curves produced by MAGIC are notably asymmetric. The recoveries estimated by the model are initially rapid, but become progressively slower. The model shows a decline of sulfate concentrations to below pre-acidification levels by year 220, and Stage 5 lasts about 100 years. Desorption of the SO_4^{2-} accumulated in the soil takes approximately twice as long as the initial adsorption.

Stage 6 Recovery of soil base saturation

After the anions have returned to their pre-acidification levels, there is a further lag in the recovery of stream base cation concentrations and ANC. Until the soil base saturation has returned to pre-acidification levels, recovery of stream base cation concentrations and ANC will not be complete. The recovery process appears to take approximately twice as long as the initial acidification. Recovery time for this hypothetical catchment is 150–200 years, only 40 years of which are shown in Figure 6.1.

Stage 7 Return to pre-acidification steady state

This stage is achieved after all ions have reached their pre-acidification concentrations.

6.2.2 Acidification time-scales

MAGIC is a simple quantitative model based on the physical and chemical processes in catchment soils. It estimates the orders of magnitude of surface water quality response times to changes in atmospheric deposition. The model shows the following results:

- Soils with moderate to high sulfate adsorption capacity and low base saturation produce acid stream water in a few decades to a century.
- Soils with low sulfate adsorption capacity and low base saturation produce acid stream water in a few years to a decade.
- Soils with high base saturation or high rates of base cation supply (through mineral weathering or artificial application, for example in agricultural soils) may buffer streams against acidification for centuries or longer, regardless of sulfate adsorption capacity.
- In response to decreased deposition, *initial* recovery rates are relatively rapid compared with initial acidification rates (i.e. the acidification and recovery responses are asymmetric).
- Total recovery times are longer (by a factor of two or more) than total acidification times.

6.3 Applied model

In more practical applications, the actual history of acid deposition at a site (or group of sites) can be used with the acidification models to examine the course of acidification at each site. The model can also be used at each site to predict its future acidification (or recovery) in response to proposed changes in future acid deposition. Such applied modelling studies of future acidification responses are an integral part of the decision-making process that leads to recommended emissions controls, and to legislation to achieve the recommended levels.

The Centre for Ecology and Hydrology (CEH) at Wallingford, UK, produced a report in 2001 that evaluated the likely response of surface waters in sensitive regions of the UK (Figure 5.4) to the emission reductions agreed upon for the Gothenburg Protocol (Section 5.3.2). The MAGIC model was used in the study to simulate the responses of groups of lakes across the sensitive regions to the acid deposition reductions that are assumed to occur as a result of the Gothenburg Protocol.

MAGIC was run for six acid-sensitive regions of the UK: the Cairngorms, Galloway, the Lake District, the South Pennines, Wales and Dartmoor. Overall, the modelling exercise suggested that the deposition reductions mandated in the emission reduction protocols should reduce the proportion of sites considered with mean ANC below zero from 21% to 4% by 2050.

Appendix 2 presents the executive summary of the report.

6.4 Confidence and uncertainty

Simulation models are useful tools for projecting the potential future response of terrestrial and aquatic resources to assumed or anticipated ecosystem perturbations (such as changes in land use, atmospheric deposition, and climate). Such projections are commonly used to integrate or synthesize scientific understanding of the dynamics of natural ecosystems. They are also often used as the basis or justification for public policy and legislation concerning management of natural resources. A continuing concern in using models, either to summarize our current knowledge or to assist in making policy decisions, is the level of confidence that can be placed in the model predictions. In a philosophical sense, models of environmental systems can never be demonstrated to be absolutely 'true'. In a practical sense, however, as a model is repeatedly tested against observation and/or experiment and found to produce satisfactory results, confidence in the model increases and the continued use of the model for either scientific or managerial activities is justified.

The model confirmation process is seldom successful at every step, however. As new or more extensive data become available, one or more aspects of the original model structure may be found wanting even as the overall performance of the model continues to be adequate. It is then appropriate to refine the model to include or improve the simulation of the missing or misrepresented phenomena. In the case of failure of the model when applied to a novel situation, the whole mathematical structure (and conceptual basis) of the model may be called into question. Most models of natural systems, however, are formulated only after extensive observation and experience with the system in question (i.e. few natural system models are built from 'first principles') and/or must be calibrated using empirical observations of the dynamics of the system (i.e. few natural system processes can be described by 'universal constants'). As a result, most mathematical models are sufficiently well constrained by the behaviour of the real system they represent that structural failures are rare and usually occur early in the development of a model.

Robust and reliable natural system models thus emerge from a confirmation process that is cyclic and progressive and that resembles the scientific method. A model structure is developed that is based on observed behaviour of a natural system. The model is tested against further (new) observations. If differences between simulated and observed behaviour are acceptable, the model is judged adequate and left unchanged. If the differences are unacceptable, the model structure is refined to improve its performance. In either case ('it worked' or 'it's been fixed'), confidence in the model is increased which leads to more applications and, in turn, leads to further refinement and/or confirmation. Models that successfully pass through several iterations of this procedure become the workhorses of scientific and managerial applications. The demise of such a model does not usually occur because of an inherent and suddenly discovered flaw; rather such models pass out of favour because new developments make possible more detailed and/or explicit models, or because the questions the model was designed to address no longer have scientific or policy relevance.

6.5 Summary of Section 6

1. Mathematical models of the acidification process that are based on physical and chemical processes in soils have been useful in scientific studies investigating the effects of acid deposition.
2. Acidification models can be used in simulation experiments to answer general questions about the complex interactions among soil processes.
3. Acidification models can be used to answer policy-related questions about expected responses to proposed future deposition levels.
4. Confidence in model predictions is gained through extensive application, testing, and revision (if necessary) of the model under a variety of conditions.

Learning outcomes for Topic 8 *Acid Rain*

After working through this topic you should be able to:

1. State the major chemical components of anthropogenic acid rain.
2. Outline the major sources of sulfur to the atmosphere and compare the magnitude of natural sources of sulfur to anthropogenic sources. (*Questions 2.1 and 2.2*)
3. Perform calculations on biogeochemical cycles, including those that determine the reservoir size, input/output flux or residence time of a compound, given two of these three variables. (*Question 2.2*)
4. Explain the effect of atmospheric carbon dioxide on the pH of 'natural' precipitation, and describe other natural factors that can influence the pH of rain.
5. Summarize, in general terms, how SO_2 is transformed to H_2SO_4 in wet- and dry-phase reactions.
6. Recognize the environmental conditions that are likely to enhance the amount or concentration of acid deposition on vegetation and soils. (*Question 3.1*)
7. Outline the major ways in which acid rain is buffered by soils through interactions with organic matter, ions derived from weathering, cation exchange reactions, and sulfate adsorption reactions. (*Questions 4.1–4.3*)
8. Define the acid neutralizing capacity of a surface water and describe the major chemical components that make up ANC. (*Question 4.2*)
9. Describe the importance of dissolved inorganic aluminium in acidification of soil and surface water. (*Questions 4.1 and 4.2*)
10. Explain the role of mobile anions in catchment and surface water acidification.
11. Explain the importance of catchment geology in determining the sensitivity of a stream or river to acidification by acid precipitation. (*Questions 4.1 and 4.3*)
12. Outline the three hypotheses that were first proposed to explain the observed acidification of surface waters.
13. Explain how palaeolimnology was used to settle some of the debates about the causes of surface water acidification. (*Question 5.1*)
14. Understand the major mitigation strategies to halt or reverse the effects of acidification, and the European protocols for limiting the emission of the acid precursors SO_2, NO_x, and NH_3.
15. Describe the role of models in understanding and predicting the effects of acid deposition.

Answers to questions

Question 2.1

(a) The major source of biogenic H_2S is from anaerobic saltwater wetland soils and sediments. Since wetland soils and sediments are included as terrestrial soils, these fluxes are shown as a terrestrial source in the global cycle (highlighted in Figure 2.3).

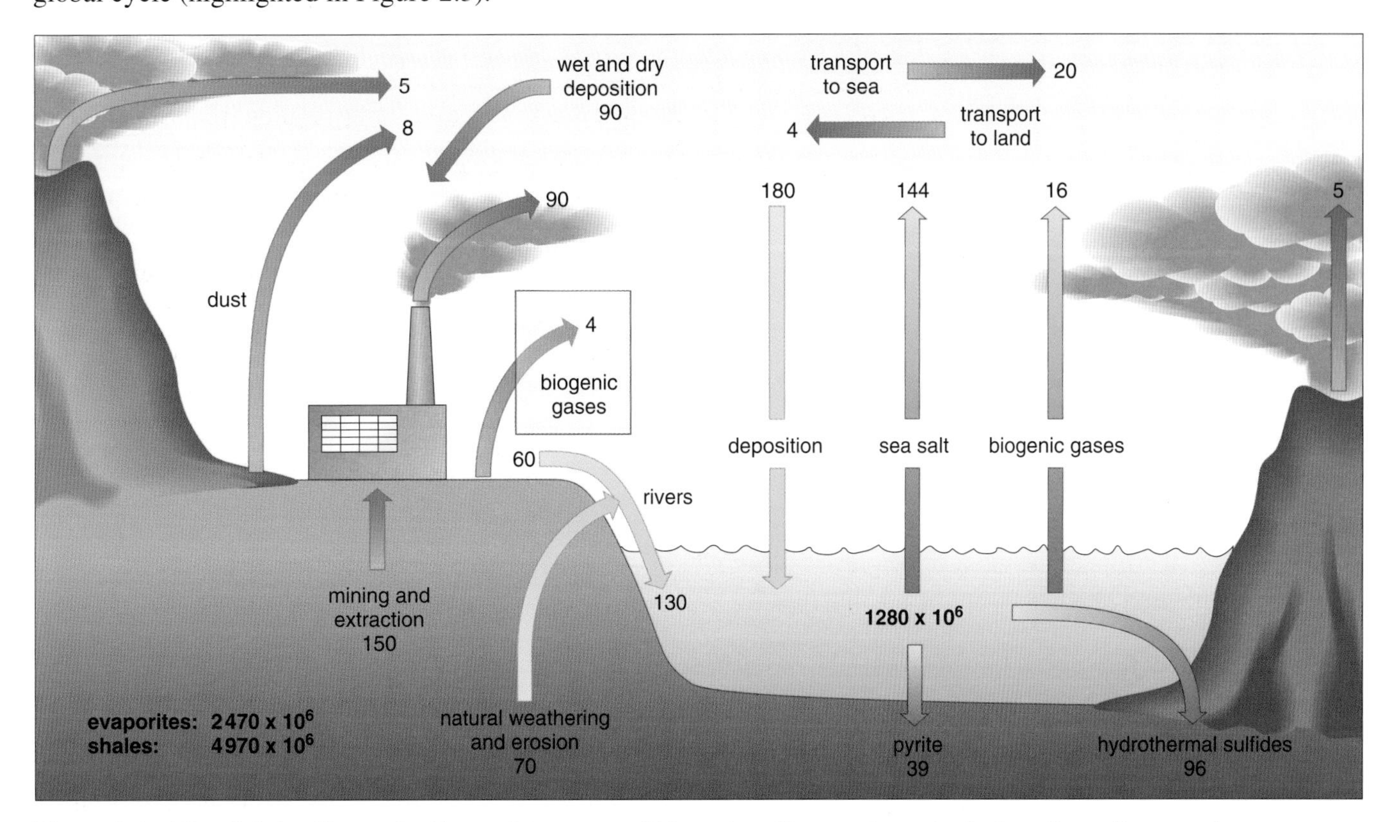

Figure 2.3 The global sulfur cycle. Terrestrial sources of biogenic sulfur are shown in the box. Input fluxes to the sea are shown as yellow arrows; output fluxes from the sea are shown as red arrows. All reservoirs (bold values) are expressed in units of 10^9 kg S and annual fluxes in units of 10^9 kg S yr^{-1}.

(b) Saltmarshes, mangrove swamps, and estuaries where anaerobic soils are exposed to high concentrations of sulfate ions are the most likely ecosystems for biological production of H_2S. In these ecosystems, the combination of anaerobic conditions, large amounts of available carbon (sediments rich in organic matter) and a large supply of SO_4^{2-} from seawater provide the ideal conditions for sulfate reduction to occur.

(c) Sulfate reduction processes remove hydrogen ions, thus neutralizing acidity:

$$CH_3COO^-(aq) + SO_4^{2-}(aq) + 3H^+(aq) = 2CO_2(aq) + 2H_2O(l) + H_2S(aq) \quad (2.1)$$

Question 2.2

(a) To balance the sulfur budget for the sea, you need to subtract all of the output fluxes of sulfur (Figure 2.3, red arrows) from all of the input fluxes (Figure 2.3, yellow arrows). The change in the reservoir size is the difference between the sum of the inputs and the sum of the outputs.

Table 2.1 Input and output fluxes.

Inputs/10^9 kg S yr^{-1}		Outputs/10^9 kg S yr^{-1}	
rivers	130	sea salt	144
atmospheric deposition	180	biogenic gases	16
		pyrite formation	39
		hydrothermal sulfides	96
totals	310		295

Change in reservoir size =

$$(310 \times 10^9 - 295 \times 10^9) \text{ kg S yr}^{-1} = 15 \times 10^9 \text{ kg S yr}^{-1}.$$

Figure 2.1 implies that the sulfur content of the oceans is increasing by 15×10^9 kg S yr^{-1}.

This value is only a few per cent of the inputs and outputs, and small errors in the input or output estimates would eliminate this difference. Many of these input and output estimates have a degree of uncertainty much higher than a few per cent.

(b) Residence time is calculated as follows:

$$\text{residence time} = \frac{\text{amount in reservoir}}{\text{flux into or out of reservoir}}$$

Taking the amount of sulfur in the ocean reservoir to be 1.28×10^{18} kg S, and (from a rough average of the totals in Table 2.1) the approximate flux into or out of the ocean reservoir to be 300×10^9 kg S yr^{-1}, or 3×10^{11} kg S yr^{-1}:

$$\text{residence time} = \frac{1.28 \times 10^{18} \text{ kg S}}{3 \times 10^{11} \text{ kg S yr}^{-1}}$$

$$= 4.3 \times 10^6 \text{ years or 4.3 million years.}$$

This implies that the ocean is a long-term sink for atmospheric sulfur, including the enhanced sulfur from acid deposition.

Question 3.1

In general, mountainous areas have higher rainfall than lowland areas, and so the sheer amount of acids deposited by wet deposition will be higher. In addition, mountain ecosystems often have more mist and fog than ecosystems downslope, and may even be shrouded in cloud. This increases the amount of acid deposition from occult deposition.

Question 4.1

In the spring, during snowmelt or when land surfaces are saturated due to low evapotranspiration, water can either percolate very rapidly through the soil, or run off the surface as overland flow. Such water has had little or no contact with soils, and therefore has had little chance to react with soil basic cations. A snowpack may store several months' worth of acid deposition, releasing it all to a stream within a few days or weeks.

Question 4.2

One reason for the healthier biota may be that the organic acids in the surface water provide significant acid neutralizing capacity:

$$\text{ANC} = [HCO_3^-] + 2[CO_3^{2-}] + [OH^-] + [org^-] - 3[Al^{3+}] - 2[Al(OH)^{2+}] - [Al(OH)_2^+] - [H^+] \quad (4.12)$$

thus neutralizing some of the H^+ and reducing the release of toxic Al^{n+} species into the surface water.

Another reason may be that much of the toxic aluminium is complexed by the organic acids in the peaty soils and never reaches the surface water (Section 4.1.5).

Question 4.3

Besides receiving more acid deposition, mountains often have resistant granitic or siliceous bedrock — their very nature as mountains means that they have not weathered as rapidly as surrounding areas. Such resistant bedrock provides only low levels of basic cations to neutralize acid deposition.

Question 5.1

If surface water acidification were a natural process, occurring over the last 10 000 years since the retreat of the last ice sheets, the abundance of the acid-tolerant diatom *Tabellaria* would probably have remained at high levels throughout the 150-year sediment core (with minor variations due, for example, to natural fluctuations in weather or other conditions), and the acid-sensitive diatom *Achnanthes* would have been absent or present at very low levels throughout the sediment core (Figure 5.7). The pH reconstruction would be constant at about 4.8 (with minor variations reflecting variations in the diatom abundances), or only decreasing very slightly, since we are only looking at a small slice of a slow 10 000-year acidification process.

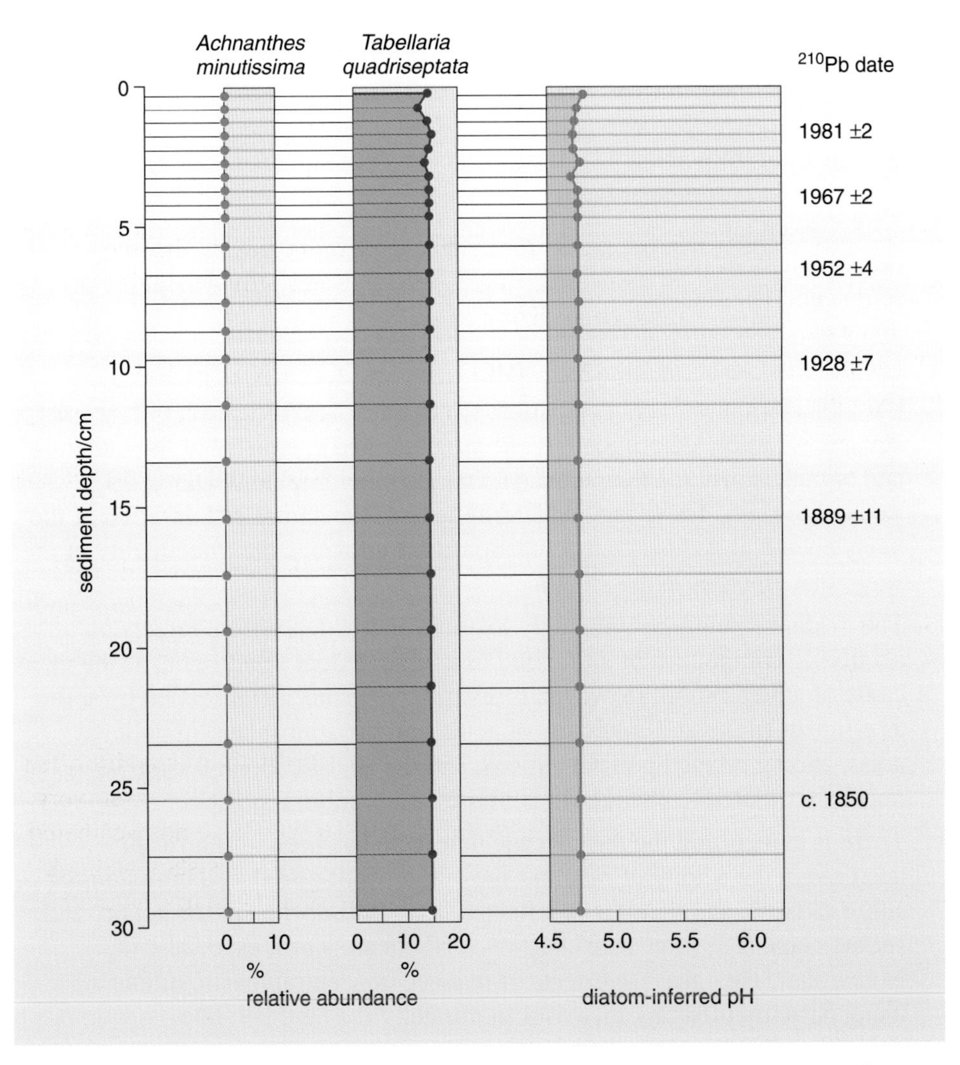

Figure 5.7 An example of relative abundance of the acid-tolerant diatom *Tabellaria* and the acid-sensitive diatom *Achnanthes* throughout the 150-year sediment core if surface water acidification were a neutral process and surface water pH declined slowly from near-neutral pH.

Appendix 1 Surface Waters Acidification Programme (SWAP)

Taken from material in Mason, B. J. (1992) *Acid Rain: its causes and its effects on inland waters,* Oxford University Press.

The main conclusions of the Surface Waters Acidification Programme are:

1. Acidified lakes and streams without, or with impoverished, fish populations occur mainly in areas that receive high levels of acid deposition from the atmosphere and have soils derived from granite or other rocks of similar composition that are resistant to weathering and low in exchangeable elements such as calcium and magnesium. Catchments with thin soils are particularly sensitive with respect to the rate and extent of acidification.
2. Examination of the remains of diatoms and other biological material in lake sediments laid down over centuries has established that many lakes in southern Norway, Sweden and in the UK have undergone progressive acidification from about 1850 until very recently. The magnitude of this acidification is appreciably greater than any that has occurred in the past 10 000 years and has marched in parallel with accelerated industrial development, as indicated by increases in several trace pollutants in the sediments. These changes and the extent of inferred acidification are geographically correlated with the intensity of acid deposition and with the geochemical status of the catchment.
3. For a given input of acid deposition, the degree of acidification of lakes and streams is largely determined by the structure and chemistry of the mineral and organic soils, and the pathways that the incoming rainwater takes through the soil. These factors determine both the nature and duration of the many chemical and biological reactions that influence the final quality of the water that emerges in the streams.
4. The evidence points convincingly to atmospheric deposition, largely of acidifying compounds of sulfur and to a lesser extent of nitrogen, as the main cause of acidification. However, forests may enhance acidification by acting as efficient filters and collectors of acid from the atmosphere in polluted areas, and by taking up metal cations, and the acidification of some lakes may be attributed to changes in land use or agricultural practice.
5. There is evidence in the past decade that there has been a significant decrease in the acidity of rain and snow as a result of reduced emissions of sulfur dioxide, and that this is reflected in a small decline in the acidity and sulfate content of some lakes. However, there are signs, especially in Norwegian lakes, that the effects of reduced concentrations of sulfate are being partially offset by increases in nitrate.

6 Fish populations, especially of salmon and trout, cannot survive in lakes and streams if the pH of the water remains below a critical level of about pH 5 for long (depending on the species and age of the fish and the chemical composition of the water). The fish are killed by the action of increased acidity and of inorganic forms of aluminium leached out of the soil by the acidified water. The effects of aluminium are ameliorated by the presence of organic acids (e.g. from peat) that complex the aluminium and render it less toxic to fish, and possibly the presence of calcium in sufficiently high concentration. However, in regression analyses based on a survey of over 1000 lakes in southern Norway in which fourteen variables in the regressions were studied, most of the variance in fishery status could be accounted for by pH, inorganic aluminium, and altitude.

7 Fish are very vulnerable to the short, sharp episodes of high acidity and aluminium that occur in streams following heavy rains or snow melt. In these episodes much of the water flows through acid soils where it is enriched in available aluminium but spends relatively little or no time in the deeper layers where it would be neutralized.

8 From carefully controlled laboratory experiments and intensive field studies, it is now possible to relate fish survival to the concentrations of acid, aluminium, and calcium in the water and to estimate the likely toxic effects of acidic episodes of differing severity, frequency, and duration. The detailed mechanisms of fish death are complex. Some forms of aluminium and H^+ ions inhibit sodium uptake and increase sodium loss, so reducing body sodium content leading, eventually, to circulatory failure. The deleterious effects of inorganic aluminium can be largely counteracted if calcium is present in sufficient concentrations.

9 Acidification and release of aluminium also lead to changes in the populations of micro-organisms, lower plants, and aquatic invertebrates. The effects of such changes in the ecosystem can include the availability of food for some life stages of brown trout and other fish.

10 The possibilities of the recovery of streams and lakes depend on the long-term balance between the catchment input and output of cations, such as Ca^{2+} and Mg^{2+}, which are exchangeable for H^+ ions. The main input of these cations in the affected parts of Scandinavia and the UK is normally by chemical weathering; the supply by atmospheric deposition is much less than that of acidifying substances. The direct cause of acidification of lakes and streams is the excess of anions of strong acids, as sulfate reduction and denitrification play only a minor role in most of the affected ecosystems. The long-term resistance of a catchment area is therefore closely related to the release of cations such as Ca^{2+}, Mg^{2+}, Na^+, and K^+. In regions covered by the last glaciation, these cations are produced mainly by weathering of minerals. Considerable progress has been made within SWAP in the determination of weathering rates as a function of mineral species, particle size, pH, production of organic ligands in the ecosystem, and the history of the soil. Estimates by different methods agree in most cases within a factor of 2 or 3. It appears that even a reduction by 60% of acid deposition would not be enough to create steady-state conditions suitable for fish in those areas that are most strongly acidified.

11 The rate at which streams and lakes will recover in response to reduced emission and deposition of acidic substances will also depend on such factors as the residence time of water in groundwater and lakes, the release of sulfate from earlier deposition, which is retained in the soil, and on changes in the land use within the catchment. In thin soils with little storage of sulfur compounds, recovery may be quite rapid. In deeper soils containing large accumulated stores of sulfur compounds, it may take several years or even decades for this to be leached out and recovery may be much slower. Recovery or restoration may be aided by liming the catchment, but this may have undesirable effects such as increased nitrification.

12 There is evidence of increased nitrate deposition but this has been only partly reflected by the increase in its concentration in surface waters, mainly because of uptake by vegetation. As the system has limited storage capacity, an additional burden of acidification could develop over years.

Appendix 2 Executive summary: *Freshwater Acidification and Recovery in the United Kingdom*

Evans, C.J., Jenkins, A., Helliwell, R., Ferrier, R. and Collins, R (2001) *Freshwater Acidification and Recovery in the United Kingdom*, prepared by the Centre for Ecology and Hydrology, Wallingford, Oxfordshire and the Macaulay Institute, Aberdeen, Scotland.

Acidification of soils and surface waters across Europe and North America results from the emission, transport and deposition of oxides of sulfur and oxidized and reduced forms of nitrogen. Acidification occurs in regions where the geology is acid-sensitive and in regions that have received a large accumulated flux of sulfur and nitrogen deposition. International efforts to decrease acidic emissions and reverse the acidification process have led to agreement and legislation within the UN-ECE Convention on Long-Range Transboundary Air Pollution and the European Union, respectively. The most recent UN-ECE agreement, the Multi-Pollutant Multi-Effect Protocol, was signed in Gothenburg in 1999 and was aimed at limiting emission of sulfur, nitrogen oxides, ammonia and volatile organic compounds by the year 2010.

The study uses the dynamic hydrochemical model, MAGIC (Model of Acidification of Groundwater In Catchments), calibrated to six acidified and acid-sensitive regions in the UK to:

(i) determine the recovery in surface water chemistry that might be expected under the Gothenburg Protocol by 2050;

(ii) compare the predicted recovery with that predicted to occur under existing agreements (Reference scenario);

(iii) compare these with the regional water chemistry predicted if no further action was taken to limit emission beyond present-day levels.

The model has been calibrated to six acid-sensitive regions of the UK — the Cairngorms, Galloway, the Lake District, the South Pennines, Wales and Dartmoor — using the best available data describing catchment soils, water chemistry, deposition and land-use characteristics.

Predictions are made using deposition fields derived from the HARM [Hull Acid Rain Model] atmospheric transport and deposition model.

In response to the deposition reductions under the Gothenburg Protocol the model predicts that by 2050 only 2% of sites sampled within the six sensitive areas will have mean Acid Neutralizing Capacity (ANC) below zero. This compares with a current figure of 21%. For the Reference scenario, the model predicts 4% of waters below zero ANC. If deposition were to remain at current levels, 25% of sites are predicted to be below zero ANC by 2050. It is clear, therefore, that emission reductions beyond the presently achieved levels will have a significant benefit for surface water chemistry in the UK.

The added 'benefit' of the Gothenburg Protocol over earlier agreements is less clear when the regions are considered together at a UK scale, but is evident in some regions. In the most heavily impacted region, the south Pennines, the model predicts 5% of surface waters with ANC below zero in 2050 under the Gothenburg Protocol compared with 12% under the Reference scenario.

The time-scale of chemical reversibility and the degree to which pre-acidification chemistry is recovered also varies regionally across the UK as a result of differences in regional weathering rates and relatively low soil base status, promoting a rapid response to change in acid anion input. For Wales and the South Pennines, lower weathering rates and the requirement to replenish larger soil base cation pools slows the surface water recovery. In several regions, notably Galloway, plantation forestry has a significant impact on predicted recovery in surface water chemistry whereby second rotation planting counteracts deposition reductions, leading to less improvement in water quality at forested catchments.

Significant uncertainties remain in these dynamic model applications relating to extrapolation of soils data, surface water chemistry data and definition of present and future atmospheric inputs, as well as process uncertainty in the model. In particular, the descriptions of sulfur adsorption and controls on nitrate leakage need better scientific understanding in order to model satisfactorily. In addition, there still exists considerable uncertainty in the response of freshwater biota to changes in chemistry.

Nevertheless, dynamic model outputs remain no less certain than steady-state critical loads constructed from similar data, but have the added benefit of providing a timescale for the assessment of existing emission reduction agreements, or for deriving future agreements.

Future research must aim to extend the regional coverage, reduce model uncertainties, incorporate stream systems, interface with Integrated Assessment Models and link with dynamic aquatic and terrestrial biological models.

Acknowledgements for Topic 8 *Acid Rain*

Grateful acknowledgement is made to the following sources for permission to reproduce material in this book:

Figure 1.3: © AP Photo/Peter Dejong; *Figure 1.4*: Building Research Establishment; *Figure 4.1*: Institute of Terrestrial Ecology; *Figure 5.1*: © Courtesy of UKAWMN, United Kingdom Acid Waters Monitoring Network; *Figure 5.2*: © Environmental Change Research Centre, Department of Geography, University College London; *Figure 5.5*: © Courtesy of Clogrennane Lime Limited; *Figure 5.6*: © Simon Fraser/Science Photo Library.

Parts of this topic are based on the S247 Case Study 'Sulphur, Acid Deposition and Lake Acidification' by Tim Allott and Lesley Smart (1996).

Section 6.2.1 was taken from Cosby, B., Hornberger, G., Galloway, J. and Wright, R. (1985) 'Time Scales of Acidification', *Environmental Science and Technology*, vol.19, p.1144, December 1985.

Every effort has been made to trace all the copyright owners, but if any has been inadvertently overlooked, the publishers will be pleased to make the necessary arrangements at the first opportunity.

Index

Note: Entries in **bold** are key terms. Page numbers referring to information that is given only in a figure or caption are printed in *italics*.

S

T

U

V

W